Masonry

Masonry

The Taunton Press

Cover photo: Lefty Kreh

Taunton
BOOKS & VIDEOS

for fellow enthusiasts

10 9 8 7 6 5 4 3 2
Printed in the United States of America

A FINE HOMEBUILDING Book

FINE HOMEBUILDING® is a trademark of The Taunton Press, Inc.,
registered in the U.S. Patent and Trademark Office.

The Taunton Press, Inc.
63 South Main Street
P.O. Box 5506
Newtown, Connecticut 06470-5506
e-mail: tp@taunton.com

Distributed by Publishers Group West

Library of Congress Cataloging-in-Publication Data

Masonry.
      p.   cm. — (Fine homebuilding builder's library)
      "A Fine homebuilding book" — T.p. verso.
      Includes bibliographical references (p.   ) and index.
      ISBN 1-56158-214-X
      1. Masonry.   I. Taunton Press.   II. Fine homebuilding.
III. Series.
TH5317.M3695   1997   97-10637
    693'.1 — dc21   CIP

Contents

Introduction

I READ IN A BOOK ONCE that all you need to know about laying up stones is: "one over two and two over one," which means basically that you should offset the vertical joints between courses. Since then, I've actually had the opportunity to lay up a few stones and have learned that, for me at least, the one elemental truth about stonework is: "whatever stone I just picked up isn't going to fit where I'm about to try it."

Stone masonry is like trying to assemble a jigsaw puzzle without having the box to look at so you know what picture you're trying to make. But good masons somehow have that picture in their heads. Generally speaking, they pick up a stone, and it fits where they put it. (This talent is an annoying phenomenon if you're working next to them, but something you'll appreciate if you're paying them by the hour.)

In the end, there's a lot more to know about stonework than "one over two and two over one," which is why this book exists. Collected here are 31 articles from back issues of *Fine Homebuilding* magazine. Written by professional masons—the kind with "that picture in their heads"—these articles cover nearly everything you need to know about stone and brick masonry. Read them carefully, and you'll learn to build walls, floors, fireplaces, and chimneys. And while you might not learn how to pick up the right stone every time, you will learn how to trim the wrong stone so that it fits where you want to put it.

—Kevin Ireton, editor

Guidelines for Laying Stone Walls
Master the basics and gravity will do the rest

by Joseph Kenlan

When I assess the quality of a stone wall, I look not at the stones, but at the spaces between them. That's where the knowledge and the craftsmanship of the mason is revealed. By reading the joints, I can tell how tightly the wall is tied together and how well the stones fit.

Dry-laid stone is perhaps the oldest form of building. Fieldstones stacked and shimmed together created rough walls long before recorded history. But because there is no mortar in dry-laid work, it has the disadvantage of being permeable by air and moisture. That's why its use is usually limited to retaining walls, where the passage of water is necessary, or to simple foundations where ventilation is desirable.

Dry-stack construction is nearly as old as dry-laid work. Mortar is used in dry-stack work, but it isn't visible. The mortar is trowelled in behind the stones, filling the voids between them. Dry-stack walls are laid stone-on-stone, like dry-laid walls. But because they're less permeable by air and moisture, they're the better choice for house walls and chimneys.

In my area, dry-stack chimneys built stone-on-stone and backed with clay mortar have stood for over two-hundred years. Although it's arguable that today's stronger portland-cement mortars make this meticulous fitting unnecessary, I prefer to build the old way. To Steve Magers, the mason who taught me stone, the only law of masonry is gravity. Gravity always prevails, so it's best to learn how to cooperate with it. Whether you're building a dry-laid or a dry-stacked wall, a garden wall or a chimney, the same basic guidelines apply.

The footings— Though it isn't always possible, my helpers and I like to provide a footing under every stone wall or chimney we build. A lot of the repairs we've done would have been unnecessary had there been footings in the first place. Footings distribute the weight of the stone and provide protection against frost heave. The footing must sit on solid, compacted earth below the frost line. It should be at least twice as wide as the wall it is going to carry. In most cases, local building codes spell out the depth and dimensions of the footings.

Sometimes, dry-laid stone footings are used for retaining walls instead of concrete. They should have broad, flat stones at the base to spread the weight evenly. In this type of footing, we usually lay 1 in. to 2 in. of crushed stone under the first course and below the frost line to help it drain.

String lines and story poles— When the footing is complete, the use of string lines or story poles will help keep the stone courses straight. For walls, two methods are commonly used. Batter boards are the preferred method for foundations and low walls (top photo, facing page). A line strung between them is used to indicate both the face of the wall and its top edge. By sighting between the line and a chalk line snapped directly on the footing (or the first course), you can keep the walls plumb without the constant use of a level.

For taller walls, story poles can be set up at each corner of the wall. They should be carefully plumbed and braced (bottom right photo, facing page). The poles serve as vertical sight lines. Strings can be pulled between them to

indicate the top of the wall or any other horizontal detail, such as a window opening. If there's framing above the work, two plumb lines suspended from it can also serve as vertical sight lines.

Hold the stone about ⅛ in. back from the strings to keep from pushing them out of alignment. Check the strings periodically to make sure they're straight and taut.

Examining the stone—Once the stone is on the site, you can begin to sort through it. Pick out the "specials" and store these in separate piles. For example, stones that have a true 90° angle and are of sufficient size should be set aside for use as cornerstones. It isn't necessary to go through the whole pile at this point. Specials can be set aside during the regular course of the work.

While going through the pile, examine the stone carefully. Most stone has a definite structure. The direction in which the stone splits the easiest is called the *rift* and reflects the way the stone was formed. The rift is very pronounced in slates and many sedimentary

A successful stonework style is as much a process of cooperating with stone as in executing a preconceived plan. Here (photo left), the use of sandstone makes the complex pattern-work possible. Gravity keeps the carefully bedded stone in place. Batter boards (photo top) are used for foundations and low walls. Lines are strung between them and they indicate both the top edge and the face of the wall. For taller walls (photo above), story poles are placed at each corner of the wall and are carefully plumbed and braced. Strings pulled between them indicate horizontal details such as window openings and the top of the wall. The story poles also serve as vertical sight lines. Here, a helper is sighting across the poles.

- Because a stone is more complicated to fit when it's bounded on both sides, it helps to lay the corner stones in a wall first and work toward the middle. That way, there's just one "closure" stone in each course.
- The height of a corner stone is its least important dimension. Resist the temptation to stand a stone on edge to get the maximum "corner" out of it.
- Ideally, the sides of a stone should be roughly perpendicular to the exposed face (or faces). Theoretically, when you look squarely at the face of a stone, no part of the adjacent sides should be visible. In practice, however, you will find that some deviation from this rule is acceptable. More important, the top and base of the stone should be nearly perpendicular to the face. The top should never slope toward the outside face of the work because this will encourage stones laid above it to slide out of the wall.
- A good rule of thumb is that you should be able to stand on a stone right after it's placed. If it rocks, either shim it with a stone chip or trim off the high spots. You can see the high spots by looking under the stone. The shadows between the stones indicate the places that need trimming.
- When laying stone, always try to anticipate problems you may be creating for yourself later on. In a sense, you must see the spaces you are creating as well as the stones you are laying. Once you have located a potential fit, try it in the wall to make sure it will do what you want.
- Each stone should fit snugly against the one next to it. The tops of adjacent stones should be exactly the same height. That way, you can span, or face bond, the joint with a stone in the succeeding course. If

the tops aren't even, leave one stone at least an inch or so higher than the other. That way, a stone can be inserted above the short one to make the tops flush. Virtually every stone in the wall should be bonded. The easiest way to see this bonding pattern is to look at both the face and the corners of a brick building. The regularity of the bricks makes the bond apparent.
- It's important that two stones not meet on a rising joint, with the stones inclining upward towards the place where they meet. Bonding such a joint is extremely difficult, and the joint creates a wedge that can snap a stone above it.
- Often you'll be left with a narrow space between two stones which otherwise fit well. Fill this gap with a thin stone but on the next course, treat this stone as if it were a joint. The stone above must span it completely.
- Rarely will a stone fit without trimming. If the stone needs trimming, mark it in place on the wall (photos below). I use a blue lumber crayon to mark light-colored stone or a piece of soapstone to mark darker stone. Then take the stone off the wall and set it on the banker or on the ground to trim it. Don't trim the stone while it's still on the wall because you can dislodge stones nearby and disturb the bond between stones and mortar. After the main trimming is done, take the stone back to the wall for a final fitting. If the stone was bedded in mortar before trimming, make sure to reset it with new mortar. Be sure to use eye protection when trimming stone.
- Thick walls require internal bonding, or header bonding to tie the opposite faces together. At least 20% of the stones on each side should extend deep into the wall at

regular intervals. It's best if the stones extend completely through the wall, but they may not be available. In that case, use stones that extend well into the wall, leaving enough space on the other face to lay a stable stone. On the next course, lay similar stones from the opposite face so that the tails overlap. You may also lay stones tail to tail at the same level and then lay a stone above them in the middle of the wall to tie them together.
- Sometimes it's desirable to use a shiner—a stone that doesn't project far into the wall. For example, shiners are used where the depth of the stone is limited by a pipe or other obstacle in the wall, or where a particular stone fits in well with the wall pattern. Be sure when using shiners that the surrounding stones tie well back into the wall. Avoid using too many shiners. An abundance of shiners and shims in a wall is one of the surest signs of a poor mason.
- You will occasionally come to places where nothing seems to fit. If you are stuck in one spot for a long time, move to another place and come back to that spot later.

Be careful of fingers when moving large stones. I often place sticks under a stone while setting it in order to allow room for my fingers. Then I pry up the stone and remove the sticks.
- Prepare for the final course before you actually get there so that you can top out with stones of substantial height. Use large stones and cut them down to meet the line.

In areas where thin stones are abundant, masons sometimes save the biggest cornerstones for the top and stand thin stones on edge between them—like books on a shelf—to bind the top of the wall.
—J. K.

When you need to trim a stone, first mark it in place with a lumber crayon or a piece of soapstone (photo left). Then remove it from the wall and set it on the banker or on the ground for trimming. Reposition the stone on the wall, taking care not to jar the adjacent stones out of position. The top should finish out flush with that of the adjacent stone so the joint between them can be bonded (photo right). Make sure you reset the stone with new mortar if the stone was set in mortar before trimming.

rocks, but is not so obvious in some igneous stone such as diabase or basalt, or in some metamorphic stone such as quartzite. Stone should be laid with the rift aligned horizontally, when possible. This positions the strongest part of the stone in the horizontal plane and prevents the delamination that is seen with carelessly laid sandstone and limestone. When the rift isn't obvious, you can usually find it by breaking a stone with a hammer or pitching chisel (more on these tools later). By splitting just one or two stones, you can usually determine the rift of the whole pile. Our general rule of thumb regarding the rift is that the more pronounced the rift is in the stone, the more important it is to orient it properly in the wall.

The grain is the second easiest direction to cut the stone. It runs roughly perpendicular to the plane of the rift, although in some stones it runs diagonally instead. The face of a stone (the visible side of a stone once it's in a wall) has typically been formed by a split along the grain. When working stone, try to plan your cuts to take advantage of the rift and grain. If you plan your work to go along with the stone's inclinations, you'll save a lot of effort.

Occasionally we run across freestone, which works equally well in all directions. This has its advantages, but it also makes it tough to distinguish between the rift and the grain. Since the rift represents the optimal bedding plane of a stone in a wall, you'll want to be careful to orient the freestone properly, or it could weather excessively. We also run across stone that doesn't work well in any direction. It can be laid with little regard to bedding planes. Examining the stone pile will also tell if there are obvious flaws and cracks to watch out for; a visual and hammer inspection is helpful here. Unless it's sandstone or other soft stone, a flawless stone will have a characteristic ring when struck with a hammer. A cracked stone will give off a dull thud.

Stone that has been exposed for long periods of time will often deteriorate to the point where it is no longer suitable for use. Cracking and crumbling are the primary indicators of this problem. Be careful when reusing stone from houses that have burned down. The heat from a fire will often crack and weaken the stone.

We also run across stone that has been quarried with explosives. This stone is more likely to have internal cracks than most stone and should also be checked with a hammer for soundness.

A visual inspection can also suggest ideas on how to lay up the stone. Certain types of stone naturally lend themselves to particular styles and patterns. Slates and strongly layered sedimentary stones, for example, are best laid on long horizontal lines. Freestone, or stone that presents larger, more vertical faces, allows the mason more latitude in design. A successful style is as much a process of cooperating with the stone as in executing a preconceived plan (photo, pp. 8-9).

If you are doing a lot of cutting, a banker will make it easier. The one shown above is built from plywood sandwiched between a pair of 2x2 frames and filled with sand on top. The principal tools for working stone on a banker are, left to right, a 3-lb. sledge (two styles are pictured here), a point, a flat chisel, a pitching chisel and a tracer. The cutting tools in the photo have heavy-duty carbide tips.

Tools of the trade—On small jobs, it's convenient to do your stone cutting right at the base of the wall. If you're doing a lot of cutting, a banker will make the work easier. This is simply a strong table on which to work the stone. A barrel or tree stump and a sand bag will serve in a pinch. The sand bag acts as a cushion for the stone to prevent it from sliding while you're working on it. We usually make a simple portable banker by sandwiching a piece of plywood between a pair of 2x2 frames. We set this on a stack of cinder blocks and fill it with sand (photo above).

Experience will determine the most comfortable working height for you. I prefer a banker about 6 in. below my waist for general work and a little higher for fine work. This allows me to swing the hammer downward, letting the weight of the hammer head do the work, not my arm.

The principal tools for cutting stone are a 2-lb. to 3-lb. sledge hammer used with either a pitching chisel, a flat chisel, a point or a tracer (photo above). I recommend using a hammer with a short wooden handle. Hold the hammer near the head for light work and slide your hand down the handle for heavier hitting. You'll find that a stone can be hit only a certain number of times before it breaks badly. Try to do as much work as possible with each blow of the hammer.

The pitching chisel, or set, is a wide blunt chisel used for blocking out the edges of a stone. It's driven at a right angle to the stone and, when struck hard, it quickly removes large pieces in a short time.

The flat chisel, or cold chisel, is used to clean up the rest of the stone, providing there's an indentation in the stone on which to seat the tip of the chisel. To remove large flakes, the chisel is driven hard at about a 30° to 60° angle. When used to smooth the edges or for fine final trimming, the chisel is struck lightly.

If there isn't a good seat for the flat chisel, a point is used instead. The point, which is held at nearly 90° to the stone, is also useful for smoothing out nobs and high spots. Some masons use a point almost to the ex-

clusion of the flat chisel for cleaning up stone. The choice is based largely on the type of stone the mason works. Points are the better choice for working hard stone. The point can also be used to put a decorative finish on the face, but this is time-consuming and is rarely done these days.

Flats and points should be kept sharp (I sharpen mine on a grinding wheel). When using a flat chisel or a point, angle the tip away from the face of the stone. Driving a chisel toward the face will break out sections of stone you want to keep.

The tracer is used like a brickmason's set to score and break a stone along a straight line. It is usually held perpendicular to the line and is struck sharply. On large stones, mark out the line with the tracer and finish the cut with the pitching tool or stone mason's hammer.

You may also want to add a thin-bladed mason's chisel to your collection. It's handy for splitting layered types of stone such as flagstone (see *FHB* #40, p. 49). A toothed chisel is useful for putting a decorative finish on softer stones such as sandstone.

For most work, hardware-store cold chisels and points are satisfactory. They usually cost under $10 apiece. Several tool companies carry carbide-tipped tools for continuous cutting in hard stone. These cost $35 and up, but hold up well enough to make them worth the price.

For light trimming, you'll also need a small hammer such as a brick hammer. I use a welder's scaling hammer for this purpose. It has two blade faces set at right angles to each other.

Stonemason's hammers come in two principal styles, which I call *American* and *European*. The American-style hammer, or walling hammer, has its roots in early America. The hammer resembles a large, blunt hatchet with a hammer face and ranges in weight from 3 lb. to 16 lb. The hammer face is used primarily to trim away large pieces of waste from a stone. If you need to break the stone along a straight line, use the blade as you would a tracer to score the line first.

The European-style hammer has a point instead of a blade, and its hammer face is smaller and more square than the American version. The point is used like a chisel point for cleaning up stone. A less commonly used style, called a *side hammer*, has two squared faces, sometimes with a carbide inset. The squared edges are used like a pitching chisel. *Bull sets* and *rifters* are heavy hammers shaped like a wood-splitting maul. They are held against the stone and struck with a sledge hammer. Think of bull sets as large pitching tools and rifters as large tracers.

For very large splitting, drills and wedges are commonly used, but this applies more to the actual quarrying of stone than to stone masonry (see *FHB* #35, pp. 35-37). □

Stonemason Joe Kenlan lives in Pittsboro, N. C. Photos by the author.

The north wall of the building seen from the inside as it neared completion. Concrete from the upper courses dripping down the wall will be removed after the entire wall is constructed. The interior has a uniform and plumb appearance.

Form-Based Stone Masonry

A method for constructing cast-in-place stone walls

by Richard MacMath

The skills needed to build a stone wall are simple enough for self-taught masons to get beautiful results after a short learning period. Of course, construction must be done with care, and a few rules kept in mind.

The cast-in-place method we used to build the walls shown here is probably the easiest for the novice. We've had "first timers" who were able to learn the technique in one day. Many were producing beautiful results in a few days.

This method is best suited to construction with cobblestones—round-shaped stones in random sizes. Often these stones are found in fence rows between fields and in sandy or gravel soils. For such stones, using formwork is simpler than laying a freestanding wall, especially if one wants a

Richard MacMath is a partner in Sun Structures, Ann Arbor, Mich. The wall is part of Upland Hills Ecological Awareness Center in Oxford, Mich.

durable, true, residential wall. Using formwork as a guide also makes the work go faster. The stacking and leapfrogging of forms ensures plumb and level results.

The job's most difficult aspect, and one impossible to avoid, is getting the stones to the construction site. Once you have found an adequate supply, hauling the stone becomes a task for as many helpers as you can assemble. Although this seems like a troublesome job, remember that transporting free stone—even over a few miles—is inexpensive when compared to the cost of other building materials.

Even if you have stone and labor in abundance, it's important to be selective in the use of stone masonry. It is best suited for retaining walls and below-grade foundation walls with outside insulation covered with earth. Foundations, basements and earth-sheltered homes are appropriate applications. In such uses horizontal

and vertical steel reinforcing bar (rebar) is required to strengthen the wall against the extra pressure exerted by the earth mass. Stone is also a good choice for interiors, especially for thermal mass walls. A passive solar design rule-of-thumb calls for 1.0 to 1.5 cu. ft. of masonry exposed to direct sun for every 1.0 sq. ft. of south-facing glass.

As a note of caution, above-grade stone walls that require insulation are a problem. They must be insulated in one of two ways: either by constructing a cavity wall or by adding an insulating layer to one side of the finished solid wall. Cavity-wall construction is difficult, time consuming and not recommended for the novice. The other option—adding insulation to the interior or exterior face of a solid masonry wall—requires studs or furring strips, insulation and a finish surface (usually gypsum board) that will completely cover one side of the stone wall. In both cases it would be easier to construct a well-insulated stud bearing wall first and add the stone veneer later.

Rules of stone masonry construction—The first rule of sound wall construction is to set the stones so that gravity rather than mortar holds them in place. Each stone must be set in a firm, relatively flat bed formed by the stone below and its covering layer of mortar. I have seen large stones fall out of the wall when formwork was removed because the novice stonemason relied too much on the concrete and formwork to hold them in place. This is the most common mistake in form-based stone-masonry. Place the stones as if you were laying a dry wall; use the concrete only as joint fill. If a level, flat bed cannot be prepared for a stone, then place it so gravity forces it inward. In this way, adequate beds can be made for angled and rounded stones.

The second rule is to place each stone so that its weight is distributed over at least two other stones below. This is the principle of crossing joints for maximum strength. Crossing joints ensures that the wall will work as a unit rather than as individual parts. This rule doesn't apply when setting small stones on top of large ones, but it's best to avoid a continuous joint from top to bottom in a wall. Otherwise you invite cracks, and the finished appearance lacks the random overlapping pattern that makes stone masonry so attractive. In long walls, however, a continuous joint is constructed intentionally, every 30 ft. or so. This is called a control joint, and it allows for expansion and contraction of the wall under changing thermal conditions.

The third rule concerns both structure and aesthetics. The mason must not only look at each stone carefully, but also at how each stone fits into place and contributes to the structural integrity of the wall as a whole. For example, a particular stone may have one face that is flat

and colorful. The beginner's impulse is to expose this face on the finish side of the wall. Unfortunately, its best use may be as a flat surface, providing a firm bed for succeeding stones. In this case, aesthetics must yield to the structural requirements of the wall.

Remember that these rules are easy to overlook when employing this cast-in-place method because the formwork holds the wall in place while the concrete is setting, hiding the finished surface from view.

Designing the wall—If the stone wall is going to be a bearing wall or a retaining wall, then the structural loads must be calculated. These loads will determine wall thickness, footing size and the amount of steel reinforcement required. Fortunately, we have engineering training so we perform the necessary calculations ourselves. Also, construction details must be figured out and drawn to scale to illustrate connections to other structural elements. For example, anchoring the roof to the top of the wall requires proper planning. Our design called for 12-in. long anchor bolts set into the top of the wall on 4-ft. centers. These secured a top plate that served as a nailing surface for 2x12 roof rafters. Since we were constructing a sod roof, we had to continue

the waterproof membrane down the outside of the wall and over a layer of rigid insulation.

In our building the north enclosing wall is a below-grade bearing wall that supports both roof and earth loads. Vertical loading from floors and roof is easily transferred to the footings because a masonry wall is very strong in compression. However, the major structural load below grade is the lateral force of the earth pushing on one side of the wall. These lateral loads force the wall to react in tension on the interior side. Pure masonry walls have almost no tensile strength, so steel reinforcement must be added. As a rule-of-thumb, earth backfilled up to 6 ft. high against a 12-in. thick wall requires no vertical reinforcing. We designed our north wall to support the lateral load of almost 10 ft. of earth. The vertical loads from the roof and the weight of the wall itself provide some resistance to the force of the earth, but we still had to add substantial reinforcement. Note that the rebar is placed on the tension side of the wall—in this case the inside—and additional vertical reinforcing was added to the bottom half of the wall where the earth loads are greatest. Horizontal rebar is laid down between stone courses at 2-ft. intervals during wall construction. The amount of reinforcement is determined by structural calculations.

Footings are generally twice as wide as the wall is thick and equal in depth to the thickness. Here, in a retaining wall situation, the wall is placed close to the interior side of the footing so that the weight of the earth on the outside helps to keep the wall from overturning.

There is one place in our building where the wall changes from a retaining wall enclosing the building to a freestanding exterior wall. Because these two sections of wall experience different loading and temperature conditions, we built a continuous control joint in between them. At the control joint the two walls are tied together with horizontal rebars wrapped with felt paper. The

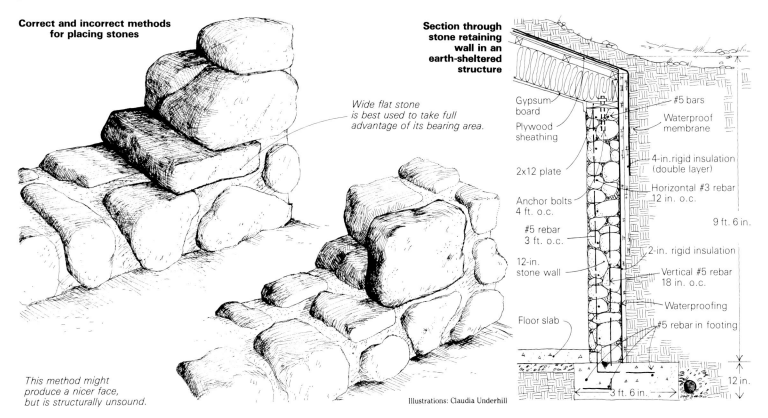

Correct and incorrect methods for placing stones

Wide flat stone is best used to take full advantage of its bearing area.

This method might produce a nicer face, but is structurally unsound.

Section through stone retaining wall in an earth-sheltered structure

Gypsum board
Plywood sheathing
2x12 plate
Anchor bolts 4 ft. o.c.
#5 rebar 3 ft. o.c.
12-in. stone wall
Floor slab

#5 bars
Waterproof membrane
4-in. rigid insulation (double layer)
Horizontal #3 rebar 12 in. o.c.
9 ft. 6 in.
2-in. rigid insulation
Vertical #5 rebar 18 in. o.c.
Waterproofing
#5 rebar in footing
3 ft. 6 in.
12 in.

Illustrations: Claudia Underhill

Steel form

½-in. pipe bolted on
90° elbow
Union
½-in. pipe
$\frac{7}{16}$-in. turnbuckle (10 in. - 16 in.)
Wood spacer
10 or 12-gauge galvanized sheet metal
4 ft.
Holes drilled for form tie wires
Metal rod for coupling to next form

Cross section of steel form on wall

Spacer
Wire form tie
8 in.
4 in.
Rebar
Footing

Wooden form

1x3 spacer
½-in. plywood
12 in.
4 ft.
2x4 frame

To add a line of forms, simply nail them together at the ends.

Cross section of wooden form on wall

Spacer
#5 rebar
Wire form tie
Key
Brace (shaded) nailed to form
Mortar bed
Footing

Wood forms work just as well as metal ones. Concrete, stones and water must always be on hand. Working with at least three or four people at a time keeps the work progressing quickly. Here two people are setting stones in a bed of concrete, layer by layer. The wood forms shown are heavier than metal ones, but last longer and are easier to clean.

rebar holds the retaining wall in place laterally while allowing for expansion and contraction in a direction parallel to the wall.

Formwork—We used both metal and wood forms, both serving the same purpose. When set on the footing or clamped to the top course of the wall, the forms held the concrete and stone in place while drying. Metal forms were lighter and easier to assemble, but sometimes bulged at the bottom as they were being filled. Wood forms, though heavier and clumsier, better resisted the weight of concrete and stone. Both types were coated inside with used motor oil to keep them from sticking to the concrete.

Design for the metal forms came from Ken Kern's book, *The Owner Builder's Guide to Stone Masonry* (Owner Builder Publications, Box 550, Oakhurst, Calif. 93644). Kern uses galvanized pipe, pipe fittings and sheet metal, together with a metal turnbuckle and wooden spacers. The forms are 12 in. high and 4 ft. long and can be interlocked at each end with other forms. We assembled 4 forms, each weighing 20 lb., but they gave us some problems. First of all, we had difficulty keeping them clean. Concrete was hard to remove from the threads on the turnbuckle and from the pipe fittings. Also, tightening the turnbuckle sometimes bent the pipe.

Generally we preferred the wooden forms even though they were heavier by about 10 lb. They were made in two parts and connected when clamped to the wall. One advantage of using wood was that the forms could be nailed to stakes and other forms for alignment and stability. We always used the wood forms when building the curved north wall. Bending pipe to the proper radius seemed difficult to us. Spacers and wire were used with these forms as with the metal ones.

Required materials and tools—Besides the stones and forms, you'll need the following materials and tools to build a form-based stone wall:

Sand, gravel, portland cement.	Mortar boards
Water	Hose, containers of water (for soaking stones)
Reinforcing steel (rebar)	
Waterproof coating (asphalt or cement base)	Wood stakes (for layout work)
Polyethylene sheets	String
Used motor oil	Gloves (lots of pairs)
Cement mixer or deep wheelbarrow and mason's hoe (has holes in it)	Wire
	Hacksaw
	Level (preferably 4-ft. length)
Wheelbarrows (for hauling stone)	Chalk line
	Paintbrush
Shovels	Wire brush
Buckets or pails	Plumb bob
Mason's trowels: large size (for filling forms with concrete), small size (called pointing trowels, for finish work)	Nails
	Hammers
	Sledgehammer
	Mason's hammer and chisels (if you want to shape stone)

We used a standard 3-2-1 concrete mix: 3 parts gravel, 2 parts sand and 1 part portland cement. To estimate the amount, assume that the concrete mix makes up approximately one third of the wall's volume. Whenever we were asked how much water to add to the dry concrete mix, my friend Wayne would reply, "to taste." There are so many variables—weather, wet or dry sand, gravel size—that the right amount of water will often vary from day to day. We preferred to use a mix that was dry enough to stand on a trowel, but wet enough to fill all the gaps between the stones. A dry mix can be held in by the form, but must be scooped out of the wheelbarrow with a trowel or shovel.

Reinforcing steel (rebar) is available in diameters from $\frac{1}{4}$ in. (#2) to $1\frac{1}{4}$ in. (#10). After calculating the amount required, you have a choice of using a small number of large diameter rods or large number of small-diameter rods. Small-diameter rebar is easier to bend and cut to the required length. For example, we often substituted three #3 bars ($\frac{3}{8}$ in. diameter) for one #5 bar ($\frac{5}{8}$ in. diameter). The cross-sectional area is the same, but the #3 bars are easier to bend into the footing and cut with a hacksaw. Since rebar is sold by the pound, there is no additional cost in using bars of smaller diameter.

Many different products are available for waterproofing below-grade walls. The standard products are asphalt-base and cement-base coatings, either of which can be applied with a brush or trowel. We brushed on an asphalt base coating and added a sheet of 6-mil polyethylene over that. Polyethylene sheets can also be used to protect the masonry during cold weather and to keep various floor and wall surfaces clean during construction.

If you're using a cement mixer (ours was driven off a wind-powered electric system), then have a few large buckets on hand for filling it quickly and easily. Using a wheelbarrow and mixing by hand requires measuring by the shovelful—a slow process. Add water by the bucketful or with a hose.

Be sure to have some large containers on hand for soaking the stones. We used old 55-gal. drums. The wall cures more uniformly when the stones are holding water as they are set into the wet concrete. Concrete and stones then dry together, and a better bond results.

Always have plenty of work gloves on hand. After the fabric of the glove is worn away, (sometimes in one day), handling the stones removes skin from your fingertips. It hurts just thinking about it. Prepare yourself for seeing a lot of useless gloves lying around with holes in the fingers.

Spend the money on a 4-ft., 3-tube level; the time saved and accuracy provided more than justify the expense. Wall surfaces will never be perfectly flat, and when you check for level and plumb, the level must span four or five stones. This isn't possible with a shorter level.

The construction sequence—Building the wall requires proper design and construction of the footings. Since we had continuous vertical reinforcement from the footings up into the wall, we placed the rebars before pouring the footings. As the concrete began to set, we chiseled a key way—a groove 1 in. wide by $1\frac{1}{2}$ in. deep—down the center of the footing. This provided a better bond between the footing and the first course of stone.

As the wall begins to go up, make sure there are enough stones stacked within reach of the

With the first course of stone and concrete in place, top, the forms have been removed. Note the width and depth of the concrete footing and the size and spacing of the reinforcing steel. The string line is a horizontal reference 4 ft. above the top of the footing supported by stakes 8 ft. high. A plumb bob is dropped from this string to plumb the wall. For a proper structural connection between the two poured layers, the top of each layer is left rough, with some stones and reinforcing steel projecting up. Each stone must be placed so that a relatively flat bed is created for the stones above. After the second course is laid, above, the metal forms are still in place on the projecting retaining wall to the left rear. The concrete has already been scraped away from these stones with a trowel. Each stone is set so that its weight is distributed over at least two stones below. Unless a control joint is required, joints are never placed directly above each other because this leads to cracks.

55-gal. drums, which should be convenient to the section of wall being worked on. One or two people should mix concrete while others fill the drums with water and stones. At the end of the day, we refill the drums for the next day.

We often worked on two different parts of the wall at once, a team completing one course while another started the next. We find the optimum team to be two people filling the forms, with one strong assistant to lug the stones, mix and haul concrete, keep the drums filled, and point the finish side of the wall once forms are removed and shifted to another section.

After the footing has cured, use a chalk line to mark the inside and outside surfaces of the wall. The first row of forms is set on these lines. We used 8-ft. 2x4 stakes with string running the entire length of the wall as a plumb guide. Adjust the stakes so that a plumb bob held from the string touches the outside form. Using 8-ft. stakes allows you to move the string up the wall as construction progresses.

The curved part of our wall was more difficult to lay out. When we were drawing the building floor plan, we calculated the radius of the inside

of the wall. After locating the center point of this curve on the ground, we marked the footing with string and chalk. We relied on the level for keeping this part of the wall plumb.

Whether you are starting at the footing or on top of a previously completed course, you should begin by hosing down the surface and removing any loose material. This ensures a strong bond. When the surface is clean and wet, set the forms in place. For the first course, this simply means setting them down on the footing with the wooden spacers in place to make sure the form is tightened to the proper dimension. For subsequent courses, the forms have to be clamped in place.

Metal forms can be erected quickly. Hold the form in place, and tighten down the turnbuckle that is just above the metal sides. Wooden spacers help prevent over-tightening. After many setups, our forms began to bend at the vertical pipe, causing the bottom to bulge out slightly. To prevent this we drilled holes in the sheet metal and improvised wire form ties that were tightened by twisting the ends around nails on the outside of the forms. Thicker sheet-metal

first if the mix is wet, but as you pack stones and more concrete firmly against the form this waste will be minimized. Pack concrete firmly around all of the rebar, both vertical and horizontal, to bond the masonry and steel.

Fill the entire form in this fashion until the stone projects a few inches above it. This provides a better joint with the next course and makes it easier to clamp the form to the protruding rocks when it is moved. When proceeding smoothly, three people were able to fill a form 12 in. high and 4 ft. long in about half an hour.

If you build a below-grade wall as we did, each side of the form should be packed differently. For the earth side of the wall, try to achieve a smooth finish for a waterproof coating by packing the concrete in tightly. On the finish side, consider final appearance as you select and pack stones. Packing the concrete is not as critical on this side since the wall will be pointed after the forms are removed. However, you won't be able to see how the faces of the stone fit together on the finish side of the wall until the forms are removed. As part of the learning process you should start the wall at an end that will not be highly visible, because until you've gained some experience you won't be able to visualize the fit of the stones.

How long the forms must be kept in place varies with the weather and the concrete mix, but they must be removed soon enough to tool the joints to a finished state. We never keep our forms in place longer than two hours, but you will have to test this timing for yourself, since the variables of mix, weather and stone may differ. Detach the metal forms by loosening the turnbuckle and removing the nails from the wire end. Pull the wire through the wall and lift off the form. Removing the wood forms requires only pulling out the wire at the bottom, since we keep the top wood strips in place. Scrape the form clean, give it another coat of oil, and set it again atop the wall. The forms need be hosed down only at the end of the day.

Once the wall surface is exposed, it begins to dry quickly, so the joints should be tooled tight to the stone immediately. On the finish side, remove concrete around each stone with a small pointing trowel. Recessing the joints in this way exposes more of each stone, making them the prominent elements in the wall and giving it a laid masonry appearance. On hot days we keep the wall wet by hosing it down with a fine spray as we work and at the end of each day. The slower the concrete dries, the stronger it cures.

To give an idea of required construction time, it took us six weeks to complete a wall 60 ft. long and 8 ft. high. The wall cannot be thoroughly cleaned until it is completed, because concrete continues to spill down the sides during construction. We waited until the building was enclosed before sandblasting the inside of the wall. This is a quick method, but you must be equipped with gloves, goggles, proper clothing, a compressor and fine sand. Scrubbing with muriatic acid and a stiff brush is another method of cleaning that removes most of the surface dirt and excess mortar. Once the wall is clean, everyone is surprised by the beauty of the color and texture in the stone. □

The exterior of a masonry wall must be kept as smooth and clean as possible for waterproofing later. In the foreground, mortar is being applied for a smooth finish. In the background, wood forms are in place on the curved part of the wall. Temporary wood forms made of 2x12s can be used when a lot of people are willing to help and additional formwork is needed.

forms, 10 or 12 gauge instead of 14 gauge, would also prevent bulging.

Wood forms take a little more time to set up. The two sides are connected by furring strips nailed into the top and by wires threaded through the form at the bottom.

After the forms are in place, use a trowel or shovel (trowels are easier to handle) to lay approximately 1 in. of concrete on the footing or previous course. Begin a layer of stones by packing them against both sides of the form and then filling in between with small stones and rubble. Make sure each stone has a face set into the bed of concrete. Some stones may be large enough to touch both sides of the form so that center-

filling isn't necessary. Pack the stones as tightly as possible, allowing minimal space for concrete. This is the best way to save money, since cement is your most expensive construction ingredient. Once a stone is laid in place, do not move it. Movement weakens the bond. Always have stones of various sizes and shapes on hand, so that placing stone around the vertical rebar is no problem.

Once the first course of stone is complete, add another layer of concrete. We often shoveled concrete into the form, and then agitated it with a small trowel, to fill gaps between the stone. Don't worry about concrete running out of gaps at the bottom of the form. This may happen at

The Structural Stone Wall

Save your stones for pointing, bed your stones in concrete, and remember that there's no substitute for gravity

by Stephen Kennedy

Adams County, in south-central Pennsylvania, has hundreds of stone buildings, most of them built before the days of portland cement. I got started in stonemasonry by pointing up some of these old beauties, and I've always been impressed with the soundness of their stonework. After nearly two centuries, these buildings are still young. The masons who built them couldn't rely on their crude lime mortar to hold stones together. Instead, they used a far stronger glue—gravity. The stonemasonry I do today follows this old-fashioned philosophy, but fortunately I'm able to take advantage of some new materials that weren't available 150 years ago—concrete and pointing mortar.

The trademark of my stonework is spotless pointing (bottom photos, p. 18). A good pointing job keeps water from getting inside the wall and also provides a consistent background that can really show off the variety of textures, colors and shapes in a rock wall.

High-quality pointing mortar isn't cheap, but I don't use a lot of it. The jointwork extends only an inch or two into the wall. The rest—the part you can't see—is just stone and concrete. There are a number of advantages to using concrete instead of mortar to lay up a stone wall. The most obvious is cost: mortar mix is a lot more expensive than concrete, and you'd need quite a few more bags to complete a comparably sized wall. This is because the gravel (technically known as *coarse aggregate*) acts as a filler in the mix, making it go twice as far without sacrificing strength. The large stones still rely on gravity to hold their position in the wall, and the concrete ensures that there will be enough space between stones for me to point later.

With a 5-gal. bucket as a measuring unit, I make and mix the concrete on site. Two buckets of sand, three buckets of gravel, ⅔ bucket of portland cement and about ½ bucket of water (depending on the moisture content of the sand) yields a large wheelbarrow load of concrete. I mix everything together at once except

the gravel, which gets added last. The final bucket of gravel really dries up the mix, and then it's ready to use.

Concrete also gives me greater flexibility in laying up the wall than mortar would. Without the added gravel to hold adjacent stones apart, mortar tends to squish out between joints. With a stiff mix of concrete, I can keep the mix well away from the exposed face of the wall. This leaves room in the joints for my pointing mortar and cuts down the risk of messing up the faces of the stones with squeeze-out. The concrete won't compress with added weight even if it hasn't set, and this allows me to work vertically without worrying about the joints collapsing. Regular mortar mix can't stand up to much compression before it sets, so you have to work horizontally, which isn't always convenient.

The search for stones—To build a structural stone wall you need plenty of stones—about 30 tons for a 2-ft. thick wall 10 ft. high and 10 ft. long. Finding the rocks and getting them to the site is at least half the work. Both your back and your pickup truck will probably suffer for it, too. Fortunately, this is stone country. Mortarless stone walls hastily piled up by earlier generations of farmers crisscross the landscape, so few house sites are bereft of material. This isn't usually enough, though, so I end up looking on mountaintops and through dry washes and stretches of woods for the many elusive "ideal" stones that almost every job demands. I sometimes take out permits to get rocks out of state forests. In fact, when I'm in the middle of a job it's hard for me to drive down the road without scanning the countryside for rocks.

With the help of fellow mason Paul Qually, the author built the walls of this small house (above) in 1977 using locally gathered stone. Large, square-edged stones that span the full thickness of the wall were saved to build corners that look and work like rough, massive finger joints.

I hardly ever cut or dress my stones because this takes lots of extra time and because it alters the rock's naturally weathered surface. If I find a nice stone that's covered with lichen, I'll often leave the lichen exposed in the finished wall.

I look for stones that are shaped like large boxes, books, bricks and milk cartons. These and other cubic forms are far easier to stack than bowling balls, footballs, turtles and sausages. On every job I always have a few unusual stones that I want to fit into the wall somehow, but the heart of the wall should be made up of parallel-sided rocks.

Wall construction—The size and construction of the footing, or footer, for a stone wall depends a lot on wall dimensions and local soil conditions. I've dismantled old walls 2 ft. thick that had no footing to speak of. I prefer to foot a stone wall with a stone base, rather than use rebar and poured concrete. This is another method used by early stonemasons, and it's stood the test of time well.

I usually start by digging a trench about 3 ft. wide and 3 ft. deep, and fill it with large rock rubble up to just below grade level. Then I pour a broad concrete cap about 3 in. or 4 in. thick. This is where the visible stonework begins (drawing, p. 19).

Keeping a stone wall plumb is always a challenge. Inevitably, some stones will protrude beyond the plumb line while others will fall slightly short of it. If the average between the proud and shy stones is close to plumb, the wall should be plumb (and look plumb) overall.

I use a plumb bob whenever I can, but sometimes there's no place to hang the bob. An alternative to the plumb line that works well for me is a 6-ft. level upended in a bucket of sand. I often use two or more of these guide sticks, positioning them as close as possible to the work in progress. I usually align the levels with the chalklines on the footer that describe the average wall width I'm aiming for. Eyeballing my stonework off these verticals is fairly easy, and they're easy to reposition.

The stones for the first course sit in a bed of concrete laid directly on the footer (drawing, facing page). Their exposed faces should fit together with a fairly even space around them. Angular stones should be positioned so the faces slope in, toward the center of the wall. This sometimes creates depressions inside the wall, which I fill with concrete or small rocks. I often have to test-fit a rock, looking under it or lifting it up to see how solid a "print" it makes on its rock and concrete base; then I adjust with more or less concrete and reposition the rock.

I hold the concrete back from the face of the wall, so that there's room between rocks for pointing. The real structural bearing starts an inch or more back from the wall face.

Vertical joints should be staggered. And it's important to use large stones here and there that span the full thickness of the wall. Large stones are essential structural and visual elements, especially at corners.

I avoid standing narrow rocks on edge in a wall. This often sacrifices the look of a really big stone, or even hides an attractive rock face. But anti-gravity stunts don't hold up over time, even if you glue a shaky stone in place with mortar. The challenge in my kind of work is to create a nice-looking wall that will last for generations.

Pointing—Before I start pointing up the joints between rocks, I hose down the wall thoroughly. This removes loose grit and dust that would prevent the pointing mortar from adhering strongly. Once the wall has dried so that it's damp rather than wet, I can start pointing. I usually buy the best cement and the cleanest, finest sand I can find. This sounds extravagant, but it isn't, because a little pointing mortar goes a long way in my walls—the pointed joint is only one or two inches thick. I use either black or white pointing mortar. For a really white mix, you have to use white sand. A grey mix can be darkened by adding black pigment. On a Trombe wall that I built recently, the north wall face is

The heart of the wall. The ingredients are stones and a very dry, stiff mix of concrete. The stones are gravity fit with overlapping joints. Concrete and small stones are used to fill the voids between larger stones. As the wall is built, the concrete is kept back at least an inch from the face of the wall. This leaves enough space between stones for a thin, strong mix of pointing mortar.

Pointing. A rich, buttery mix of pointing mortar made with fine sand sits well on the trowel and is easy to work. At left, Kennedy uses a pointing trowel to work the mortar into joints between stones. When the mortar starts to dry, he goes over it with a stiff, dry brush, right, smoothing the joint tight against the stones and brushing away small splatters and crumbs. Pointing mortar should be kept damp to prolong its curing time.

Photos: Dannie Kennedy

pointed with white mortar, while the south-facing joints (photo center right) are black for better solar absorption.

I always use a rich pointing mix: 2 parts fine sand, 1 part cement. This keeps the mortar buttery and generally easy to work; it won't slide off your trowel or out of the joint as easily as a 4-to-1 mix will.

You've also got to keep the mix as dry as possible. I add just enough water to get the mortar past the crumbly stage. This way, I can pack the joint well without having the mortar run down the face of the wall. I load a triangular mason's trowel with a fist-sized blob of mortar, flatten it and then pack it into the joint with any one of several thin pointing trowels (photo facing page, bottom left). The wider the pointing trowel, the better it will hold mortar, but for thinner joints you need skinnier trowels.

The more pointing you do, the less you'll tend to lose mortar off your trowel. A few drips are inevitable, though, and if these land on exposed rock faces, just leave them be. If you can resist the urge to clean up these splatters immediately, you'll avoid staining the face of the wall. Let the misplaced mortar stand until it's very dry, but not hard; then scrape it off with a trowel or stiff-bristled brush. It should fall off like dust.

Every so often there's an especially large or broad joint that looks out of place among its narrower neighbors. When this happens, I pack pointing mortar into the space and then push a small face stone into the mud. This fills in the space nicely and eliminates unsightly fat joints in the finished wall. Whenever I insert such non-structural stones, though, I make sure that they penetrate the full depth of the pointing mortar. Otherwise, they're liable to fall out sometime in the future.

Another way to fill fat joints is with small ornaments. Not all clients like this kind of thing, but a sculptor I built for recently supplied me with quite a nice variety of inserts, including a brass horse, a steel toy truck (photo bottom right) and a small, cast-bronze Mickey Mouse.

Once the mortar is worked into the joint, I don't fuss with its rough texture for about an hour or so. This gives the mud time to harden up slightly. I then scrape it back and smooth it with a pointing trowel. After troweling the mud smooth, I go over each joint with a small (2-in.), dry paintbrush (photo facing page, bottom right). This really tightens the joint against the edges of the stones, and it removes any remaining grit from the mortar.

The next day after pointing, when the mortar is very firm, I give the wall a thorough hosing down. This helps the mortar to cure better by prolonging its drying time. With hose in hand, you can also go over the wall and wire-brush away any stray mortar globs that might have escaped your scrutiny earlier. The second day after pointing, I wet the wall again, this time setting the nozzle to its hardest spray. This is the last chance to brush off excess mortar without having to resort to muriatic acid or sandblasting. Once the pointing mix has cured, what's on the rock will pretty much stay there. □

Stephen Kennedy lives in Orrtanna, Pa.

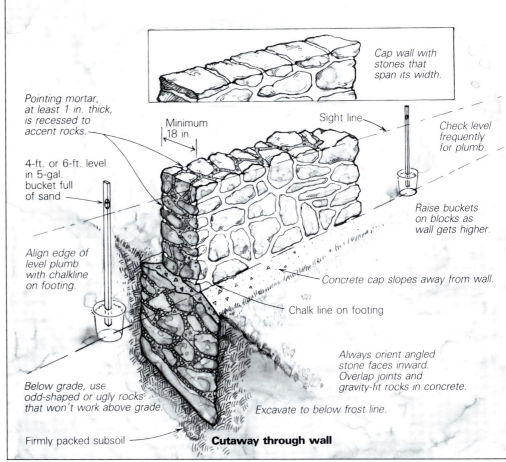

Pointing mortar, at least 1 in. thick, is recessed to accent rocks.

Cap wall with stones that span its width.

Minimum 18 in.

Sight line

Check level frequently for plumb.

4-ft. or 6-ft. level in 5-gal. bucket full of sand

Raise buckets on blocks as wall gets higher.

Align edge of level plumb with chalkline on footing.

Concrete cap slopes away from wall.

Chalk line on footing

Always orient angled stone faces inward. Overlap joints and gravity-fit rocks in concrete.

Below grade, use odd-shaped or ugly rocks that won't work above grade.

Excavate to below frost line.

Firmly packed subsoil

Cutaway through wall

Keeping the wall plumb and strong. *Instead of a conventional concrete footing, the author frequently fills a trench with a rubble of rock, gravel and concrete. The trench should be excavated twice as wide as the wall to a packed base below the frost line. The rubble footing is topped with a concrete cap 4 in. to 6 in. thick that slopes slightly away from the wall. Stones should be gravity fit with overlapping joints. Some stones will have to protrude slightly beyond the intended width of the wall, while others will set just shy of it. If plumb lines can't be set conveniently, you can work off sightlines established by two or more levels upended in buckets of sand. The levels should be positioned at opposite ends of the wall, vertically aligned with the layout line chalked on the footing.*

Skill, patience and a varied selection of shapes and sizes make it possible to accomplish intricate stone joinery without cutting any stones to size. Kennedy built the wall shown above in his own house and pointed the joints with dark mortar. Several stones have a natural lichen mantle that Kennedy decided not to disturb. At left, a toy steel pickup truck is used to fill a large joint that would otherwise stand out badly in the finished wall. Small, non-structural stones can also be used in this way, but they must be pressed firmly in the pointing mortar before it sets.

Building a Slip-Form Wall

Free materials and a novel approach shape a stone cottage

by Doug Miller

Stonework stands apart from most construction techniques. Laying stone is slow work, which affords you the time to observe and enjoy the construction as it progresses. And instead of making trips to the lumberyard for more materials, you sort through piles of irregularly shaped stones, often found nearby. There's no cashier to pay.

As a builder in northern Michigan's Keweenaw Peninsula, I use native stone as a building material whenever I can. So when it came time for my wife and I to build our own cottage, we quickly settled on stone for the foundation and walls. As for the framing, sheathing, paneling and flooring, we elected to use lumber straight out of the neighboring forests, milling much of it ourselves. We worked without electricity, which enabled us to hone our hand-tool skills while building a comfortable cottage with a kitchen and living room downstairs and two loft-like bedrooms upstairs (top photo, facing page).

The setting—The site is located down a three-mile-long logging road and is surrounded by a maple forest broken in places by fractured basalt cliffs and ancient bedrock bluffs bearing glacial scratches. At the base of the cliffs, basalt rocks and boulders are plentiful and are scattered as randomly as babies' blocks. The tan-colored basalt is very heavy and has a weathered, sandpaper-like surface. Angular cleavages make it ideal for laying in a wall. It's not easy to get, though, because it involves working on unsteady talus, and more often than not, the truck is far away. Nevertheless, a large portion of the cottage would be built from these rocks.

To provide contrast to the rough-surfaced basalt, we decided to mix in some smooth-surfaced basalt, too, which is grey-black in color. I gleaned it from local streams and from the nearby shores of Lake Superior in early spring, right after winter had reshuffled the deck. We also blended in granites from the lake and streams, red conglomerate from old mine shafts and Jacobsville sandstone, an orange-red and white stone, which we rescued from old buildings and quarries.

Amassing the stones—Gathering the stone for the cottage was nothing short of an expedition. I went on all my forays armed with a 6-ft., thin-pointed pry bar to lift the stones loose, a 10-lb. stone hammer to dislodge stones and to test them for cracks, and a wheelbarrow. At this stage, I was extra careful to select suitable stones so I wouldn't end up hauling a lot of stones we couldn't use.

Most of the stones we found were crumbly, or they had hairline cracks or undesirable shapes. Visual inspection located most of the cracks, but not all of them. Wet rocks were especially tricky because the cracks were hard to see. When in doubt, I employed the age-old test of rapping a stone with a hammer and judging the stone's utility by the sound it makes. I was also careful to pick stones that presented a flat face to the exterior, with the sides and top roughly perpendicular to the face (for more on stonework, see "Guidelines for Laying Stone Walls" on pp. 8-11).

To lug the stones to the job site, I used the bed of my "heavy half-ton" pickup and a double-axle trailer. In all, I hauled almost 100 tons of stone. About 90% of the stones would end up in the walls of the house. The extra 10% gave us a good selection of stones to work with.

As I unloaded the stones at the job site, I sorted them into separate piles according to size and shape, laying them face up. This filing system greatly reduced clutter and minimized long searches for the right stone. As the stonework progressed, I went through the piles and reorganized the stones whenever I felt I wasn't seeing the best possible selection.

Down in the trenches—Instead of hiring a backhoe to excavate the foundation trenches for the cottage, I decided to dig them by hand. I dug the trench 44 in. deep, which was all the way down to the bedrock, and 32 in. wide. The job took nine days, but it was reassuring to know that our heavyweight building would rest on a solid base.

The sidewalls of the trench were stable, permitting a 14-in. by 32-in. basalt-rubble footing to be layered up form free. The footing consists of alternate layers of concrete and basalt (drawing, p. 22). Vertical lengths of ½-in. rebar spaced 3 ft. o. c. stick out the top 18 in. and were later tied to vertical rebar in the foundation wall. The rebar eventually extended all the way up to the top of the wall, with joints lapped a minimum of 18 in. Other than a single vertical length of rebar on either side of the windows and doors and a length

at each corner, this was the only rebar used in the wall. A level layer of concrete caps the footing, with a 1½-in. by 1½-in. keyway pressed into the top of it to further engage the foundation wall.

Once the footings were cured, I built forms on top of them for 18-in. thick foundation walls. The forms consist of 2x4s and 1-in. thick pine boards, with used motor oil brushed on the insides to serve as a form release. I bolted and tie-wired the forms together and braced them with earth fill and 2x4 kickers.

I poured the foundation wall up to grade level, filling the forms about half with stones and half with concrete. That's just enough concrete to fill all the voids and secure the stones.

During the pour, I tapped the backs of the forms frequently with a hammer to work the concrete down into the voids between the stones. Once the pour was completed and the forms removed, I sloped a mortar bed on the exterior ledge between the footing and the foundation wall to direct any water seepage away from the footing. That done, I backfilled the trench.

A hybrid wall—Now I turned my attention to the walls above grade. Stone walls can be either laid or formed. I think a laid wall usually looks better than a formed wall because the mason deliberately places each stone and can clearly see the outcome. A formed wall is quicker to build and is just as strong as a laid wall, but the stones are sandwiched between forms, giving the mason less control over the appearance of the finished wall. For the cottage walls, we decided to combine the two methods.

I laid the stone on the exterior side of the wall and formed and poured a 50:50 mix of concrete and rubble on the interior side, finishing the interior later with insulation and pine paneling. This system allowed me to lay several courses of stone around the exterior in a visually pleasing pattern, then erect forms on the inside and pour the concrete-and-rubble mix between the wall and the forms. I repeated this process up to the top of the wall, using keyways to tie adjacent pours together.

The bottom of the wall about is 18 in. wide, the same width as the foundation wall,

20

but the rest of the wall is 14 in. wide. This created a 4-in. concrete ledge a couple of feet above grade to support the floor framing (drawing p. 22). The fireplace chimneys on both ends of the cottage were built of used brick woven right into the stonework (bottom photo).

Forming a strategy—Instead of building new forms for the walls, I overhauled the foundation forms, cutting them into 2-ft. by 8-ft. panels. For the 14-in. thick portion of the wall, I figured out a system to erect the forms that made use of about 150 threaded bolts ⅝-in. by 9-in. long that I had salvaged earlier (drawing, p. 23).

Once I laid the first few courses of stone in the 14-in. wall, I drilled a pair of ⅝-in. holes at the top and bottom of each form, spacing the top holes the same distance from each other as the bottom holes. Then I propped the forms into position, bolted their ends together and plumbed and braced them with 2x4s nailed to stakes driven in the ground. I also tied the forms to the laid wall with wire ties embedded in the wall. Next, I poured the concrete and rubble. When I reached the top holes in the forms, I poked the 9-in. bolts through the holes and embedded them into the concrete, then finished the pour and laid a few more courses of stone on the exterior.

When it came time to remove the forms, I lifted them up and hung them on the same bolts, this time sticking the bolts through the bottom holes in the forms. When I reached the top holes during the next pour, I again stuck bolts through them into the concrete in order to support the forms for the next course up. It took five pours to complete the wall. Once the walls were finished, I cut the bolts flush with the inside surface of the wall with a hacksaw and snipped off the wire ties.

Working the mortar—I mixed the concrete for the formed walls with an old 1-hp mixer. I mixed the mortar for the laid walls by hand in a wheelbarrow using 1 part masonry cement to 3 parts clean semi-coarse sand. Before adding water, I mixed the dry ingredients together with a hoe. Then I mixed in small amounts of water until I had a rather stiff, elastic mortar. The water content of mortar is critical, so I was careful not to add too much.

Before laying a stone, I dampened the stone and its bed with a small paintbrush, dipped in a can of water, for optimum mortar-to-stone bonding. Then I loaded up my mortar board (called a hawk) with mortar from the wheelbarrow and troweled it onto the bed a few inches back from the face of the wall. I set the new stone into the mortar, allowing the weight of the stone to squeeze out the excess mortar. I carefully removed

The entry of the cottage is flanked by salvaged glass blocks (photo top). The roof is a mosaic of fiberglass shingles, with galvanized sheet metal at the eaves to prevent the buildup of ice and snow. The walls are built of smooth and rough-textured basalt, with granite, conglomerates and sandstone blended in for added color. The mix of small and large stones adds to the visual interest (photo center). Bricks salvaged for the fireplace chimneys were laid up at the same time as the walls and were woven into the stonework (photo bottom).

the excess with a small trowel and returned it to the hawk. I then moved on to the next stone, returning after the mortar joint had set up a bit to point it with a ⅜-in. pointing trowel. To create a thin shadowline, I recessed the joints about an inch.

I removed any mortar left on the rock face or edges with a wire brush (I used a fine-bristled pot-and-pan variety) after the mortar became firm so it wouldn't streak the stone.

Anytime I nudged a bedded stone, I either tapped it down or reset it in new mortar. At the end of each day, I scratched the fresh joints lightly with a wire brush, imparting a rock-like surface to the incompletely set mortar.

Tools and string lines—Most of the stone I used was hard and tough, so I shaped the stones only when I had to. You don't need a whole shop full of tools to do stonework. I got my tools from an old gravestone company, including a variety of chipping and splitting hammers up to 12 lb. and chisels and points of every shape and size. For this job I used a 16-oz. chipping hammer for nipping the small corners and dressing down the shims. When I split or shaped a bigger stone with chisels or points, I reached most often for my 6-lb. stone hammer because it's the easiest to control.

I used story poles and string lines to ensure that the exterior face of the wall was laid plumb and that the doors and windows were positioned at the proper elevations. For the story poles, I nailed vertical 2x4s to each of the batter boards, checking to make sure they were plumb and that their outboard edges lined up with the building lines. That allowed me to stretch a pair of string lines between the poles and visually align the face of each stone with the strings. To locate horizontal features such as window sills and door headers, I drove nails along the edges of the story poles at the proper elevations and stretched strings between the nails.

Getting a lift—Many of the stones I used for the cottage were in the 75-lb. to 125-lb. range, not exactly feather pillows. I was very careful to lift with my legs and not my back, and I often passed up a nice rock because it was too heavy to lift. During the entire wall construction, I was fortunate to get only a few pinched fingers and an occasional stubbed toe.

To hoist the large square-angled cornerstones, I used a come-along and a pulley hung from a tripod made of ironwood poles. It worked well, but next time I'll use a chain fall instead of a come-along. A chain fall is a ratcheted system with a chain and pulleys, often used for lifting engine

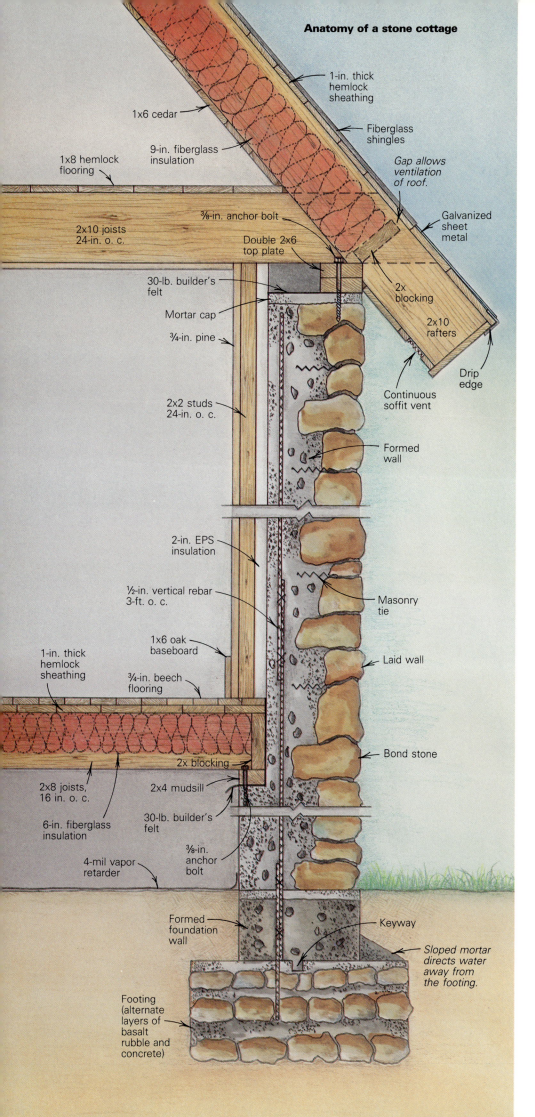

Anatomy of a stone cottage

1-in. thick hemlock sheathing

1x6 cedar

Fiberglass shingles

9-in. fiberglass insulation

Gap allows ventilation of roof.

Galvanized sheet metal

1x8 hemlock flooring

3/8-in. anchor bolt

2x10 joists 24-in. o. c.

Double 2x6 top plate

30-lb. builder's felt

2x blocking

2x10 rafters

Mortar cap

3/4-in. pine

Drip edge

Continuous soffit vent

2x2 studs 24-in. o. c.

Formed wall

2-in. EPS insulation

1/2-in. vertical rebar 3-ft. o. c.

Masonry tie

1x6 oak baseboard

Laid wall

1-in. thick hemlock sheathing

3/4-in. beech flooring

2x blocking

Bond stone

2x8 joists, 16 in. o. c.

2x4 mudsill

6-in. fiberglass insulation

30-lb. builder's felt

4-mil vapor retarder

3/8-in. anchor bolt

Formed foundation wall

Keyway

Sloped mortar directs water away from the footing.

Footing (alternate layers of basalt rubble and concrete)

---

blocks. The chain has a hook on one end, which is attached to the object to be lifted. The lifting is done by tugging on the free end of the chain. It's a cousin to the come-along, but it's a lot faster.

One stone in particular gave me a scare. A perfect 600-lb. cornerstone has no business being in the top of a wall, but this one measured up perfectly and had a 3-ft. bed waiting to accept it. While cranking it up, I could hear the ropes groan while the tripod poles quivered and creaked, but soon I had the stone dangling over its preplanned site. Suddenly the ground under the tripod's back leg gave way, causing the stone to swing end for end and drop onto the wall. Looking up, I was amazed to see the stone sitting perfectly in its new position. I got back up on the wall, pried the rock up enough to pack in some mortar, lowered it back into position and took the rest of the afternoon off.

Bond stones—Virtually every vertical joint in a laid wall should be spanned by the stone above it, which is called face bonding. I was careful to face bond the walls of the cottage at every opportunity (middle photo, previous page). The key to face bonding is to place adjacent stones so their tops are level with each other, providing a stable bed for the stone on top. When that isn't practical, small stones or shims can be used to bring the tops up to level. In addition to face bonding the walls, I laid in plenty of bondstones that span the width of the wall, tying the laid wall and the formed wall together. Some codes call for at least one of these bondstones every 3 ft., both horizontally and vertically, and that's the standard I adhered to. For insurance, I tied the walls together with 7-in. long masonry ties spaced about 2 ft. o. c. horizontally and 1 ft. o. c. vertically.

Window and door frames—To create the openings for the windows and doors, pressure-treated wood jambs were braced into position and the walls were laid and poured around them. Lag bolts screwed into the backsides of the jambs anchored them to the concrete. For headers, I cut cedar timbers from a nearby swamp and dimensioned them into 8-in. high by 14-in. deep beams with a portable sawmill (Wood-Mizer Products, Inc., 8180 W. 10th St., Indianapolis, Ind. 46214-2430). The ends of the headers were moored to the concrete with 2-ft. long anchor bolts made out of 3/4-in. threaded rod. Wood in contact with concrete can rot, so I soaked the ends of the headers with wood preservative and stapled 30-lb. felt over them.

The final course of stone on top of the wall was the most time-consuming to lay because all the tops had to be roughly level. The extra supply of stones was a big help here, giving me a wider variety of shapes and sizes of stones to choose from. Once the stonework was completed, I capped the walls with a level bed of mortar.

Drawings: Michael Mandarano

Raising the roof—After all that heavy lifting, the remaining work seemed easy. With friend and carpenter Mike Clack pitching in, the two of us quickly framed the roof with 2x10 rough-sawn spruce rafters spaced 24 in. o. c., nailing the rafters to double 2x6 top plates bolted to the tops of the walls with ⅜-in. anchor bolts. The rafter pairs were tied together near the ridge with 2x10 collar ties and at the walls with 2x10 floor joists. The finished upstairs floor is 1x8 hemlock. We installed a skylight to illuminate the stairwell, then decked the roof with 1-in. thick hemlock sheathing.

To finish off the roof, I used 20-year fiberglass shingles. I have often admired the older, multicolored shingles that have become practically obsolete on today's market, so I worked out a multicolored pattern on graph paper and bought a few squares of each color. The resulting roof echoes the colors of the stone in the walls and resembles the shadows cast by the tall maples of the surrounding forest. I also installed an 18-in. wide strip of galvanized sheet metal on the eaves in order to prevent the buildup of ice and snow.

A simple venting system—The roof is insulated between the rafters and between the collar ties with 9-in. (R-30) fiberglass batts. Because we used 10-in. rafters, this left a 1-in. airspace between the insulation and the sheathing. This afforded me the opportunity to use a simple, inexpensive roof-venting system. Continuous soffit vents allow air to enter the roof at the eaves and to travel between the insulation and the sheathing until it enters the attic space above the collar ties. The hot air then exhausts through triangular vents at the gable ends.

Filling in the floor—The next step was to frame the first floor. First I laid 30-lb. felt on the concrete ledge cast into the interior of the wall and bolted a 2x4 pressure-treated mudsill over it with ⅜-in. by 6-in. anchor bolts embedded in the concrete. Then I installed 2x8 spruce floor joists 16-in. o. c., nailed solid blocking between them and decked the joists with 1-in. thick hemlock boards. I finished up by laying down beech flooring diagonally over the hemlock.

The floor is insulated between the joists with 6-in. (R-19) fiberglass insulation held in place with wood lath, and a 4-mil polyethylene vapor retarder covers the earth in the crawl space. Vents built into the stone wall keep the crawl space fresh. For each vent, I omitted a rock in the exterior of the wall and blocked off the opening inside the form before pouring the interior. Then during the pour, I embedded ⅛-in. hardware cloth in the concrete midway through the opening to keep out insects. I like the looks of these vents a lot better than the standard aluminum foundation vents.

Warm walls—A stone wall is inherently porous, wicking both air and moisture from the outside of the building to the inside. To stop air and moisture from entering the house, I insulated the interior side of the walls with 2-in. thick (R-10) EPS insulation.

Once I cut the insulation to fit around the window and door openings, I stuck it to the wall with a few dabs of construction adhesive and then framed 2x2 stud walls up against the insulation, spacing the studs 24 in. o. c. Between the studs, I installed 2x2 blocking 24 in. o. c. to serve as nailers for wood paneling. Instead of fastening the stud walls directly to the concrete, the top and bottom plates are nailed to the ceiling and floor joists. This prevents any fasteners from conducting heat out of the building. The walls add 2-in. of dead air space to boost the R-value of the walls a bit.

The stairwell provides all the necessary interior partitions both downstairs and upstairs. I milled all the interior paneling a couple of years earlier, so the wood was well-seasoned. I paneled the downstairs with white pine, one upstairs bedroom with cedar and the other upstairs bedroom with aspen, all local woods. I trimmed out the skylight with maple because it has good reflective qualities, and I made the baseboards out of walnut and red oak. I was surprised and a bit pleased when the cottage was completed to realize that not a single sheet of plywood had been used. □

Builder Doug Miller lives in Calumet, Michigan. Barbara Miller, his wife, co-authored this article and lent a skilled hand to the building project more than once. Photos by Bernie O'Brien.

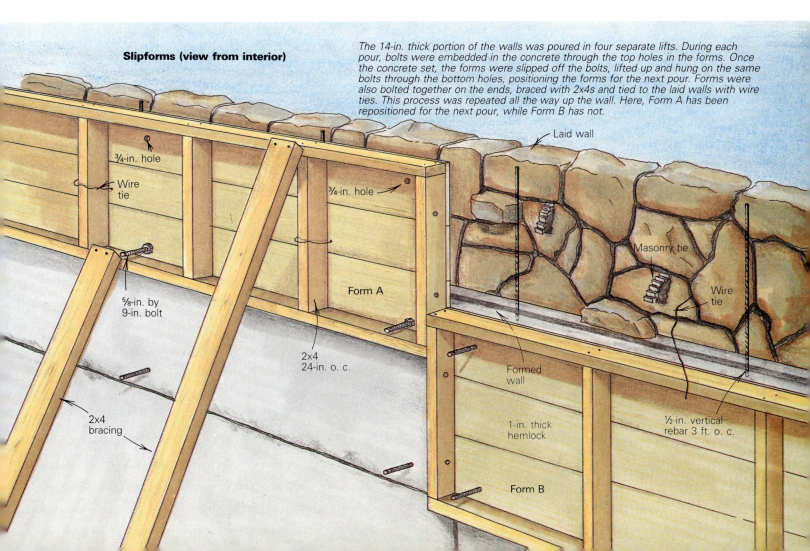

Slipforms (view from interior)

The 14-in. thick portion of the walls was poured in four separate lifts. During each pour, bolts were embedded in the concrete through the top holes in the forms. Once the concrete set, the forms were slipped off the bolts, lifted up and hung on the same bolts through the bottom holes, positioning the forms for the next pour. Forms were also bolted together on the ends, braced with 2x4s and tied to the laid walls with wire ties. This process was repeated all the way up the wall. Here, Form A has been repositioned for the next pour, while Form B has not.

¾-in. hole

Wire tie

⅝-in. by 9-in. bolt

2x4 24-in. o. c.

2x4 bracing

Laid wall

¾-in. hole

Form A

Masonry tie

Wire tie

Formed wall

1-in. thick hemlock

½-in. vertical rebar 3 ft. o. c.

Form B

Laying a Granite-Faced Wall

Techniques for using stone that is scarce and tough to shape

by David Tousain

Although Iowa offers some of the world's finest farmland, it does not abound in laying-quality stone. Several years ago, Charles Carpenter and I began collecting stone to build a foundation for a solar house. The time it took to collect stones and the immutably slow pace of dressing the stone eventually caused us to give up on "getting done this season." After six stone-laying seasons, we had completed a 36-ft. by 42-ft. full basement. In the beginning, we approached our work with two rules. First, use the stones as found—no shaping. Second, the 18-in. thick walls would be 100% stone—no filler. These rules changed as we became more skilled and sought the look of tighter work, and as it became evident that there would not be enough laying-quality stone for the whole job.

The all-stone wall changed during construction to a tightly laid face of stone on the inside backed by mortared stone rubble and 4-in. by 8-in. by 16-in. solid concrete blocks (bottom drawing facing page). We tied the blocks to the rubble with corrugated masonry ties. To provide structural bonding between the wythes, we wove large stone and block headers through the rubble to bond the two faces.

We later parged the block and waterproofed it with trowelable bentonite and insulated it with rigid foam. The bentonite, which we ordered in 5-gal. buckets, performed well for much of the job. Some batches, however, were of poor quality and wouldn't hold fast to the wall. Dry bentonite panels might have been less troublesome. At the entrance to the walkout basement we laid stone outside and insulated the wall inside (photo right).

Gathering, cleaning and sorting—Most of the loose stone in the area is a fine-grained granite in a pleasing variety of solid colors, burgundy and purple being the most common. A few miles to the north is the limit of the Wisconsin glacier, the last to cover this area 15,000 years ago. As it moved south, it pushed the granite, an igneous rock, and some metamorphic rocks ahead and left in its wake a rounder, softer stone. We scavenged only the stone south of the limit, pulling it from roadsides and fieldstone piles. Farmers were happy to let us inspect their collections and we were careful to rebuild piles as we hunted through them.

Unlike softer stones and masonry products, granite has a low porosity, which gives it relatively poor ability to bond with mortar. A sim-

At the archway entrance to the walkout basement, the stone swaps places with the concrete block to appear on the outside.

ple hosing-down can't remove bond-destroying lichen and baked-on dirt, so we thoroughly scrubbed each stone with brush and water (bottom left photo, p. 26).

After washing the stones, we sorted through the collection for size and shape. It took a lot of grading to find stones of usable size and shape, with reasonably flat faces and with top and base surfaces that would provide structural support. To ease the stone-hunting task, we built tables from 2xs, plywood and block. These allowed us to categorize stones in order of increasing height and to place them with best faces forward at a "shopping" level that saved time and energy. The tables couldn't handle all the stone, so we still had to search through the unsized pile.

From natural to shaped stone—The more we followed rule #1 (use the stones as found),

the more we thought, "This one would really be useful if we could just knock this nob off." As we became more discriminating in our work, we did begin shaping stones with cold chisels. It became clear that we needed more substantial tools, so next season we bought a 3-lb. hammer and carbide-tipped chisels. These chisels are expensive, but nothing else would stand up to hours of pounding granite. A ¾-in. flat chisel and a ⅞-in. dia. point made an adequate combination for each of us and would last about three weeks' worth of hard work before we had to replace them. (Our chisels came from Bicknell Manufacturing Co., P. O. Box 627, Rockland, Me. 04841).

We soon set a goal of using ⅜-in. mortar joints throughout. This meant that most stones had to be marked and cut to fit. Depending on its size, a stone was either hand-held or shimmed to rough orientation, using eye or torpedo level, and marked to show bottom and side cuts. Red pencil worked well for marking—it is easy to see and does not get knocked off by the first few hammer blows. It was then a matter of sitting on a sturdy bench with stone between the knees—wearing a face shield and a leather farrier's apron—and shaping the stone with chisel, point and hammer. A strip of rubber tire nailed to the working area of the bench was a good shock absorber as well as a bench protector. When a stone was too large to fit between the knees, we shaped it in a banker—a wood box filled with pea gravel and mounted on concrete block at a comfortable workbench height.

To work the stone, we firmly scribed a line at the red mark with a flat chisel and sledge. We'd strike more forcefully as this line became deeper, but not so much as to reduce the stone to rubble (this occasionally happened despite our best efforts). It takes about 60,000 years for Mother Nature to erode an inch of granite, so this stone is not about to yield easily to hammer and chisel. We came to know a given stone's workability by its color, texture and density, although there were always a few surprises. Blue stones were heavy and unworkable, dark peach stones were grainier and softer. Some stones were so hard that we simply laid them as found or saved them for landscaping.

Checking for fit—After shaping a stone, we checked it for fit, fine-tuning it if necessary. Large stones generally had flat top and bottom surfaces and needed little work to true them. Some were irregular or rounded, however, and looked awkward sitting on a bed of flat-topped stones with shims filling in around them. Instead of shimming them, we made cardboard templates of the irregular stones, and used them as a guide to prepare a conforming bed of smaller, but still good-sized, stones (bottom right photo, next page).

Determining which stone went where was a mixture of accommodating sizes, adhering to good stone-masonry practice and sticking to our personal aesthetics. The tension between these factors inspired creativity and comprised the largest part of a day's work. On some days we'd find the stone right at our feet, while other days demanded endless searching for a rare proportion. Because progress was slow, we had plenty of time to improve our skills. The walls are a visual record of this change, evolving from a rough rubble style to a more tightly laid style, called coursed and roughly-squared. Looks did not rule over proper techniques, such as using stone in its most stable position and keeping top and bottom surfaces fairly level through the thickness of the wall, possibly slanting down into the wall but never out (see "Guidelines for Laying Stone Walls" on pp. 8-11).

As we focused on shaping and laying individual stones, it was important to keep in mind the overall picture of plumb walls with carefully placed wall openings. We accom-

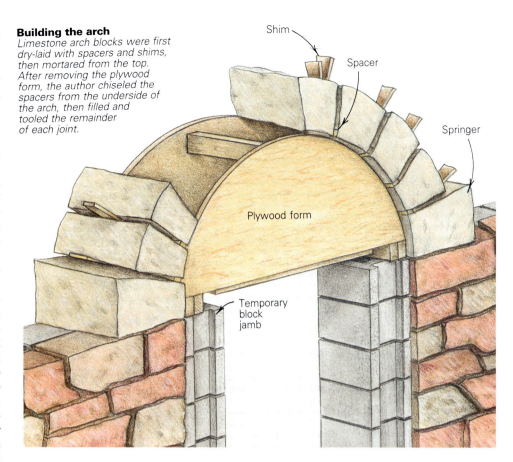

Building the arch
Limestone arch blocks were first dry-laid with spacers and shims, then mortared from the top. After removing the plywood form, the author chiseled the spacers from the underside of the arch, then filled and tooled the remainder of each joint.

Shim

Spacer

Springer

Plywood form

Temporary block jamb

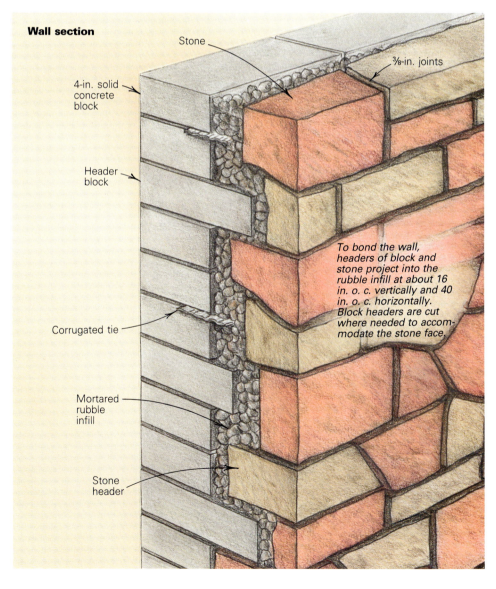

Wall section

Stone

⅜-in. joints

4-in. solid concrete block

Header block

Corrugated tie

Mortared rubble infill

Stone header

To bond the wall, headers of block and stone project into the rubble infill at about 16 in. o. c. vertically and 40 in. o. c. horizontally. Block headers are cut where needed to accommodate the stone face.

manual reprinted by Dover Publications (31 E. 2nd St., Mineola, N. Y. 11501). Type-M mortar is recommended for walls below grade. It contains a 1:1:6 mixture of portland cement, masonry cement (a mill-mixed mortar that includes lime) and sand. We reduced this to four parts sand, giving us a more plastic mortar that was easier to handle, had better water retention and was slightly stickier for use with the nonporous granite.

Usually, we laid no more than two courses a session. Going higher created the risk of putting too much pressure on fresh mortar joints. Because of the irregularity of the stone and occasional tight joints, we found that a ¼-in. steel woodworking chisel was best for working joints: the tip was good for routing while the beveled backside was effective at tooling. After everyday's laying, we cleaned up the edges of the joints with a stiff nylon brush and water from a squeeze bottle. This was easier than cleaning off dried mortar later.

In the heat of the summer, when fresh mortar joints can turn to dust in a couple of hours, we learned to cover our work with old, white blankets. First we used a dry blanket to reflect heat until the joints were ready to tool; then we used sopping wet blankets covered with polyethylene to keep new work damp until the following morning. We held down the poly and blankets with lengths of angle iron bent at right angles. With this attention, joints always turned out rock hard. After each section of block and stone had set, we would fill the cavity with Type M mortar and stone rubble.

Building an arch—We built two arches with salvaged blocks of limestone (top drawing, p. 25). After setting the springers—the two blocks whose inclined tops form the springline of the arch—we stacked up temporary jambs of concrete block. We built a plywood and 1x centering, or form, for the arch and set it on the blocks, leveling it with shims and bracing it with 2xs. We then dry-laid the arch blocks with shims and ½-in. plywood spacers between the blocks. After spacing all joints equally, we used trowels and push sticks to fill the joints with a wet, rich mortar. We tooled the joints, kept the arch moist for twenty-four hours, then removed the form. The spacers in the underside of the arch were then chiseled out of the joints, and the joints were filled and tooled. We kept the arch moist and wrapped in polyethylene for two more days.

As the walls slowly grew, we had time to salvage materials for the rest of the house, such as 8x12 fir beams (found in time for us to form the beam pockets), full 3-in. by 12-in. fir joists, and unused 50-year-old clay tiles, which will visually tie the roof to the earth-colored foundation. We may not take so much time closing in the house, but we refuse to hurry. □

David Tousain lives in Coon Rapids, Iowa. Photos by the author.

The 18-in. basement walls are faced on the inside with granite and backed with rubble infill and concrete block. Over six seasons of stone-laying, a random pattern (at right in the top photo) was replaced by a semi-coursed pattern that required precise shaping. Rather than shim under a large stone with small stones, the author made a cardboard template of the stone, then shaped a conforming bed of smaller stones (photo above). To give the nonporous granite the best grip on mortar, lichen and dirt were cleaned from the stones before shaping (photo left). Family members helped out with that tedious task.

plished this by setting and bracing salvaged concrete forms on the footings. Using a transit, we shot elevations of beam pockets, window sills and the top of the wall, then snapped lines on the forms. We laid all the walls with the block face abutting the forms. By continually squaring from these lines and measuring down to the top of the last course, we were able to plan ahead and end up with good-size stones in the top course. Measur-ing to the form was also a simple way to maintain consistent thickness in the wall and keep its inside face plumb.

Mortar—After dry-fitting enough stone for a day's laying, we'd mix the mortar. We looked at mortar as a bearing agent, not a bonding agent that is supposed to defy gravity. For help in selecting a mortar type, we consulted *Concrete Masonry and Brickwork*, an Army

Facing a Block Wall with Stone

A good rock supply, tight joints and hidden mortar are the secret to a solid, structural look

by Tim Snyder

Building with fieldstone and building with concrete block represent two extremes in masonry construction. Concrete blocks aren't especially interesting to look at, but they go up fast, and it's easy to build a sturdy wall with them. Stone construction demands patience, skill, and above all, lots of rocks. Even with these ingredients, the different shapes and sizes of the material make it tough to keep a wall of stone plumb and strong. Given these considerations, it's easy to appreciate a construction technique that combines the beauty of stone with the strength and practicality of concrete block.

Larry Neufeld laid up his first stone face to cover a block chimney in a house that he and his brother were building. He had never worked with stone before, but as a general contractor he knew enough about masonry to take on the project. By the time work began on the solar addition shown here, he had developed a technique and style that take the best from both building materials. The finished wall—20 tons of mass facing the windows and skylights on the south side of the addition—shows little mortar at all, and unless you examine the joints carefully, they seem to be dry-fit.

A flexible system—Neufeld's method uses found stone, and thanks to the New England countryside, he can usually gather what he needs from the fields and stone walls on the owner's property. Working against a 6-in. thick block wall, he lays up a face 8 in. thick, using odd-sized stones from 2 in. to 7½ in. thick. The void behind the stone is filled with mortar, which sets up around the masonry ties set in the block's joints.

Neufeld's system can work just as well with a poured wall or a bearing wood-frame wall, as long as the footing is beefed up to hold the extra load, and there is a mechanical connection between wall and face.

It's good to begin the job with plenty of

Hiding the block. **With a depth of 8 in., the stone face that covers this block wall doesn't require rocks of uniform thickness. Careful fitting is still important, though. At left, Neufeld works against temporary grounds that frame an opening in the wall. These boards were later replaced by the oak casing shown on the next page. The finished wall faces a bank of windows in a solar addition, and looks like a solid stone wall.**

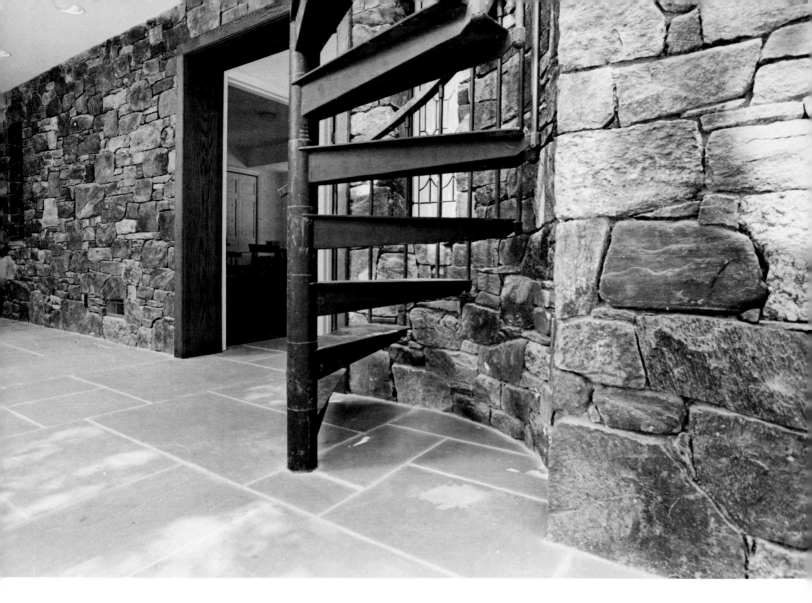

stones. As you look for rocks, pick out natural corners, base-course stones with especially flat, broad faces and pieces with unusual colors or mineral formations. Toss these in separate piles before you start building, and each time you sort through your rock, take stock of the sizes and shapes you've got. Cataloging like this can make the job go a lot more quickly. But even with a good collection of stone, you can expect to be missing a few key pieces. In the middle of a job, Neufeld often finds himself driving more slowly past stone walls after work, seeking out an elusive corner or curved face.

Before laying up the face, be sure that any wall or ceiling surfaces that will be adjacent to the stones are finished. This means drywalling, paneling, plastering and painting earlier than you normally would, but it's far easier to do this work before the face goes up.

Laying up the wall—For bonding stone to block, Neufeld uses a mix of three parts sand to one part portland cement. Working his mix in a wheelbarrow, he adds just enough water to make a very stiff mortar. Then at the front of the barrow he adds a bit more water and trowels up a small section of wet mix. The stiff mix is used between the stones so that no mortar will flow out of the joints onto the exposed rock face. The wet mix fills voids closer

to the wall, and bonds the back sides of the stones solidly to the block.

Neufeld uses a tape to check the thickness of the face as he lays it up. A level isn't much help because of the irregularities in the stone, so he uses it only for rough checking. Working against a plumb block wall is pretty good insurance that the stone face will be plumb, but with many rough facets to account for, Neufeld does plenty of adjusting by eye. Fortunately on this job, a temporary post had to be nailed up to support the second-floor overhang where the circular stair would go. By plumbing the post both ways, Neufeld was able to use it to align the face of the wall as he laid it up. He also measured against the post to check the arc of the curved wall section.

The secret to achieving the dry-stack look is to test-fit all stones carefully before laying them in place, and then to keep the mortar away from the face edge of the joint. Test-fitting the stones is the first step in building the face, and it's like working on a big jigsaw puzzle. Working horizontally across the wall, you have to find stones that fit well together. A good fit means not only that the joints are tight, but also that they are staggered vertically, just as they would be in a solid, structural wall (see Neufeld's finished wall, above).

In many instances, you have to do some coaxing to get a joint right. And it's always a

good idea to flatten the top edges of your stones slightly before casting them in the wall. This ensures a stable surface for the next course of stones to rest on.

Neufeld uses a mason's hammer to knock off leading edges, and a cold chisel to fracture thick stones. Sometimes you can split a sedimentary rock along its bedding lines, but more often than not you'll end up with random fragments. This is one reason why Neufeld prefers to trim off as little as possible, using smaller pieces to fill gaps rather than trying for an ideal fit between two stones. He doesn't like to chip into the exposed face of a stone if it can be avoided, explaining that a split face has a harsh look that will never be lost inside a house.

Your test-fit stones should be able to rest on the previous course without falling off. They don't have to be exactly plumb at this stage, but you're looking for a gravity fit. Once you're satisfied that a group of stones fits well into the wall section, memorize their relative positions and remove them from the wall. Then prepare a bed of mortar by packing some stiff mix on top of the previous course. Work out from the block wall and leave the inch of joint area closest to the outside face bare of mortar. Lay down just enough mortar so that each stone will seat securely in its preassigned position. After pressing the

Hiding the mortar. The key to the dry-fit appearance is to keep mortar away from the face edge of each joint. Neufeld packs the mortar close to the block, left, then presses the stone in place and seats it in its mortar bed with several hammer blows, as shown above.

Balance and alignment. At right, stone chips inserted along joint lines to serve as temporary wedges prevent tall, thin stones from leaning out of plumb. They're removed after the mortar has set. The facing is thick enough to conceal a heating duct in the wall. Below, corner and curve construction depends on a good selection of shapes and sizes. A mixture of small and large stones also makes it easier to stagger the joint lines.

stones into their mortar bed, Neufeld sometimes uses a hammer to help seat them.

As you seat the stones, check to make sure they're plumb. Broad, narrow stones that don't extend the full depth of the face tend to lean out farther than they should. To make minute adjustments in orienting these stones, Neufeld inserts small rock fragments that serve as temporary wedges. They hold the stones in alignment until another course is laid up and the mortar sets; then Neufeld removes them.

It's best not to pack mortar behind a course until the mortar between the stones has set. Then trowel in the wetter mix to fill the space behind the stone, and you're ready to test-fit another course. This way the wetter mix can't ooze out of the joints and dribble down the face of the stone.

At the end of the day when you're using up

the last of your mortar, don't fill all the way up behind the last course of stones. It's better to leave a slight depression because this forms a keyway for the mix you trowel in the next day. Another important practice at the end of the day is cleaning the stone you've laid up. Go over the joints with a pointing chisel or a sharp piece of wood, and rake them back so that there's little or no mortar showing. Then use a broom to sweep down the face of the wall so that any drops of mortar are removed before they adhere.

Curves, corners and openings—To build the curved section of the face, Neufeld traced the clearance arc for the circular stair onto the concrete floor. Then he fit and laid up the stones as if he were working on a straight section. The only differences were that he had to use smaller stones to get smoothly around the arc, and that he could no longer sight off the temporary post to check for plumb. He used a level instead.

Successful corners are mostly a matter of having a good variety of cornerstones to choose from. Your first inclination may be to overlook small, right-angled rocks in favor of large, squarish stones. But what you actually want is a mixture of large and small; this creates overlapping joints and integrates the corner with the rest of the wall, as shown in the photo, bottom left.

You don't need an exact 90° angle to make a cornerstone. The secret is to aim for a right-angle average over several courses. Stones that come within 10° of 90° should work in a corner, as long as you get a good combination of large and small, acute and obtuse.

At door or window openings, both the stone face and the block wall are exposed. On this job, Neufeld hid this joint with trim. Before constructing the face, he erected temporary grounds from 2x stock along the trim lines. Once the face had been built into these plumb and square housings, they were removed and replaced by finish trim.

Fine points—Neufeld admits that it takes time to develop technique and style, a consistent choice of stones that will look nice in the finished wall. He likes to play one shape off against another, but stresses that the joint lines should give an impression of horizontality. Using a mixture of small and large stones is important to the overall composition, and also makes it easier to stagger the joints. But with this method of laying up stone, you've got the flexibility to try out your own ideas. Neufeld says that on his next job, he'd like to do a rock pattern in relief.

Building this 32-ft. wall took Neufeld about 400 hours. Since he was the general contractor for the entire solar addition, working on the wall kept him at the site through the arrival and departure of most of the subcontractors. Because the structural part of the wall—the concrete block—was finished at an early stage, there was no need to rush in laying up the stone. Having the time to find and fit the rock is an important advantage. □

Stone Veneer on Concrete Block

The right mortar and care in laying stone are the keys to a beautiful job

by M. Scott Watkins

Lines of a springhouse. The author's admiration for early American springhouses helped spawn the design of this stone-veneered storage building.

As a builder and sometimes stonemason, I welcomed the chance a few years ago to design and build a small outbuilding with a lower story faced in stone veneer. Designed to capture the charm of an early American springhouse (photo facing page), this small-scale project was a break from the additions and renovations that are the mainstay for my three-man crew and me. The job gave us a chance to practice different skills. And with the help of our French-trained carpenter and mason Nicholas Lombard, the job took only three weeks from start to finish.

Barry and Virginia Wood needed a garden shed-playhouse where their six young children could store bikes, toys and sports equipment. Barry favored a stone storage shed reminiscent of the outbuildings of homesteads in rural Pennsylvania and New York. Virginia liked the more formal wood-frame outbuildings found in colonial Williamsburg, Virginia. My design blends the two. The first-story walls are stone veneer over concrete block, and the second-story walls are wood frame with hardboard siding and painted trim. Stone arches over the windows and the door, which we laid up over temporary wooden supports, add an attractive architectural detail.

Stone veneer offers a builder more design flexibility and lower costs than solid stone construction. If we had used solid stone, the 8-ft. by 11-ft. building would have required walls 16 in. thick or more for stability and to meet building codes. But walls that thick would have elevated labor costs and diminished floor space. Instead, the lower walls of the building are 6-in. stone veneer over 6-in. concrete block. Where the outside walls of the building are concealed by a timber retaining wall, we used 12-in. block and no veneer. The veneer looks almost identical to solid stone masonry on the outside (although veneer of less than 6 in. thick looks weak at the corners—you might call it stone wallpaper). The tricks to making the job come out right are the right mortar mix and careful placement of stone.

First, the block wall—Before any stone could go up, we had to construct the concrete-block inner wall for the first floor. While our apprentice, Evan Johnson, mixed mortar, stacked blocks and sorted square-edged stones for use on the corners, Lombard built the door and window bucks. Bucks are the wooden frames used to make rough openings in walls for doors or windows. We made ours from 2x8 pressure-treated southern yellow pine, rabbeting the corners and assembling the bucks with screws and waterproof glue. The bucks must be square so that windows and doors will fit. We braced ours with diagonal 1x material so that they wouldn't be distorted as we laid block, stone and mortar against them.

As Lombard and I laid up the blocks, we set, braced and anchored the bucks (top photo, this page). Over each buck we placed two steel lintels to carry the weight of the block. The L-shaped steel is ¼ in. thick and 3 in. on each side. Along each course of block, we placed corrugated metal wall ties every foot or so and inserted truss-shaped reinforcement wire in every third

Setting the bucks. After the door buck has been made and braced, it is set into place and plumbed. Note the masonry buck anchor tie that has been screwed to the left side of the buck—it will help lock the frame to the masonry walls.

Solid foundation for veneer. The galvanized, corrugated wall ties are easily visible in mortar courses between the blocks. This photograph also shows the 6-in. overhang of the 2x12 mudsill.

Bolting the mudsill. After the concrete-block walls have been placed, they are filled with mortar. Then the 2x12 mudsill can be set over the ½-in. dia. anchor bolts. Note the steel lintels over the windows.

Corners set first. The stone corners were set first, then the author and his crew worked toward the middle of each wall. Mortar was carefully raked back from the face of the wall.

A precise fit. To build the arches, stones are first dry-fitted on top of a centering, which is a temporary support. Working from a skewback (a stone forming an inclined plane) on each end of the arch, the masons then set stones in mortar. Sand placed on the centering helps keep mortar from marring the underside of the stones in the arch.

course. The corrugated wall ties we used were galvanized steel about ⅞ in. wide and 7 in. long. About one-third of the wall tie was put in the block wall, and the rest was left for the stone wall. These strips of metal are mortared into the stone veneer to tie the two walls together (bottom left photo, p. 31). We used masonry buck anchors to link window and door bucks to intersecting block and stone walls.

Once the 7-ft. high block walls had been completed, we filled the hollow cores of the blocks with pea-gravel concrete to increase the mass of the blocks. We also set ½-in. dia. anchor bolts in mortar in the top row of blocks so that we could install the treated pine mudsills (bottom right photo, p. 31)

We used 2x12 mudsills installed so that they overhang the blocks on the outside by 6 in. The overhang provided a guide for setting the stones plumb and true as the walls went up and helped make a smooth transition from stone to wood-frame construction above.

After the concrete-block walls were up, electricians installed conduits for electrical service and outlets on the outside of the block. Running electrical wires illustrates two of the advantages of veneer-over-block construction: The wire runs are easier to make than they are in solid stone construction, and when the stone veneer is applied over the block, the conduits are concealed.

Selecting the stone—Conventional wisdom discourages mixing different building stones in the same wall because it can be difficult to make a pleasing pattern of different colors and textures. But after a visit to a retail stoneyard in Sterling,

Virginia, the Woods couldn't settle on any single type of stone for the building. They did narrow the choice to three favorites, so we bought Shenandoah fieldstone for its shape and texture, Pocono gold building stone for its color and Red Oak building stone for both color and texture. Shenandoah fieldstone is a natural-weathered sandstone that is collected in Virginia. Pocono gold is a sandstone quarried in eastern Pennsylvania, and Red Oak building stone is a quarried granite from southern Virginia. We added some Pennsylvania split-weatherface building stone, mainly for corners and window openings. Pennsylvania split weatherface is a natural fieldstone collected from the surface and split into shapes suitable for building. We also sprinkled in some native quartz that we found right on the site.

The Woods commented on how best to mix the colors, sizes and textures of these different stones. They preferred a random-sized, weathered stone and told me they wanted the building to look old. Our mix was heavy on the natural fieldstone for a weathered look (some of the stones still had lichen and moss on their faces) with a sprinkling of the others for a variety of color and texture.

In all, we needed about 3½ tons of stone for the veneer work (the stone weighs about 65 lb. per cu. ft.). We ordered far more than that. The surplus gave us plenty to choose from as the walls went up and gave the Woods enough stone for the dry-stacked retaining wall, walk paths and steps they built when we were through with the outbuilding. The stone was delivered wrapped in chicken wire and stacked on pallets. The stoneyard uses a special truck equipped with a boom, which makes unloading all that stone painless. Prices probably vary around the country, but for us the most expensive stone in the lot was the Pennsylvania split weatherface at about $250 a ton.

Mixing mortar and setting stone—There are perhaps as many stone-mortar recipes as stonemasons. For neatness, strength and workability, I prefer a relatively dry, almost crumbly, mix of one part masonry cement (type H), one part portland cement (type I) and six parts masonry sand. The masonry cement is a bagged mixture of portland cement and lime for workability. The straight portland cement adds strength. This 1-1-6 mixture produces a mortar called type M by the American Society for Testing and Materials. Of all mortars tested, type M is the strongest with a 28-day compressive strength of 2,500 psi. I mix the cement and the sand with just enough water for hydration and workability. If you squeeze a handful, no water will run out, but the surface of the resulting ball will be moist. I try to avoid smearing the stone faces so that I don't have a tedious cleanup later. Substituting one part pea gravel for sand helps stabilize the stones when building up a high area such as a corner.

Although it's strong and neat, this mortar mix has two drawbacks: It doesn't fill voids readily because it's so crumbly, and it can dry out before reaching maximum strength. We carefully compacted the mortar under, behind and be-

Photos this page: Mario Pareja

tween each stone to fill any voids. We also soaked the concrete block with water each morning and gently misted the set stones several times daily to ensure adequate water for complete hydration. (Hydration is a chemical process in which a material in a plastic state becomes solid, gains strength and hardens in the presence of water.)

Setting rough stone requires a different technique than building a block or brick wall where all of the pieces are of uniform thickness and height. To keep the walls plumb, we would set a stone in place and measure from the outside face of the concrete-block wall to the outside face of the stone. We tried to keep this distance to 6 in. We also held a straightedge—an 8-ft. 2x4—between a chalkline on the slab and the outside edge of the 2x12 mudsill to check the stone placement as we went. But there's a funny thing about working with stone: Setting pieces by eye is sometimes better than using a level or a straightedge. The rough faces of a stone can sometimes look wrong, even when the level says they're right. When in doubt, trust your eye.

We began at the corners and worked toward the middle, weaving the stones together and working up a foot or so at a time, avoiding continuous vertical joints. "One stone over two" (top photo, facing page) is the rule; the wall should look as if it would stand up without mortar between the stones. Once the corners were set, we filled in between them, setting stones mostly by eyeballing from corner to corner. We measured from the face of the concrete wall, using a straightedge only when in doubt.

Ideally, mortar joints should suit the scale of the building. In this case we tried for ½-in. wide joints, though some joints are smaller, and some are wider. Many stones needed minor trimming to fit their neighbors. We used brick hammers for chipping small high spots and a chisel and a sledge in a sand box for larger cuts. Some of the stones were not suitable for hand trimming because of their grain (for more on splitting stone, see pp. 8-11), so some joints are not as tight as we would have liked. We're not looking for perfection; we want a smooth, flowing look. Raking the joints back a consistent ½ in. helps achieve a smooth look. Store-bought tuck pointing trowels worked fine for longer joints. For smoothing short and irregular joints, we used a ¼-in. wide pointing trowel cut down to about 2 in. in length and the handle of a broken cement finishing trowel shaped on a grinder. Occasional misting with water from a garden hose helped us rake the joints to a smooth, dense face.

Stone arches—The arches we built over the door and the windows make an attractive detail on the building, and constructing them was one of my favorite parts of the job. The first step is to construct the temporary formwork, called a centering. The centering supports the arch stones (the voussoirs) during construction. We had previously cut extra arch-shaped head moldings for the window and door bucks to use as centerings. We clamped these pieces to the buck heads to provide a sturdy centering at each opening (bottom photo, facing page).

Pointing the soffit joints. With the mortar set and the centering removed from the arches, the author points the mortar joints.

Finishing the top. A conventional wood-frame second story completes the building. The structure is sided with beaded hardboard.

The next step was to lay up stones to the centering and form a skewback at each intersection of wall and arch. A skewback is a stone that forms an angle, like an inclined plane or wedge, to support the thrust of the arch stones. We let this work set up overnight.

The next morning we were ready to choose and set the arch stones on the centering. Studying the pile of wedge-shaped stones set aside and saved for the arches, I held a mental image of the completed arch and sorted out some 20 or so likely candidates. I assembled these stones in my mind in various combinations before trying them on the centering. I tried out the fit and decided on a combination that looked balanced and pleasing.

With the arch stones evenly spaced dry on the centering, I removed three or four stones and re-set them with mortar. I worked my way across from each skewback toward the center, leaving the keystones (stones at the center of the arch) and the adjoining stones for last. To prevent excess mortar from marring the undersurface (soffit) of the arch, I squeezed a bed of moist sand onto the centering between each stone before packing the joints with mortar. Sand was also useful under some stones to keep them from shifting as I worked. After allowing the arches to set overnight, I removed the centering and cleaned and pointed the soffit joints (top photo, this page). The second-story playhouse (bottom photo, this page) was framed, finished and roofed conventionally with 2x4 walls and 2x6 rafters. □

M. Scott Watkins is a designer-builder in Arlington, Va. Photos by the author except where noted.

Top photo, this page: Mario Pareja

Pointing Stonework

A pastry bag speeds this tedious job and gets the work done cleanly and efficiently

by Christopher Kachur

Filling the mortar bag is messy. Hold the pastry bag upright over the mortar pan so that any mortar that drips down will fall back into the pan. Use any type of heavy-duty plastic container to lift the mortar out of the pan and pour it into the bag. Fill the bag only a little more than halfway.

Get a feel for how the mortar will flow. Get a good grip on the open end of the mortar-filled bag, and twist it to create pressure on the narrow end; test it first by squeezing some mortar back into the pan to get a feel for how the mortar flows and how much pressure is needed to get the job done.

Whether you're a veteran stonemason or a motivated novice, pointing stonework with a trowel and pointer can be time-consuming, messy and tedious. But you can speed the process and get good results and a neat job by using a pastry bag instead of a trowel.

I used to point my stonework at the end of each day. However, many variables can affect the color of the mortar when you mix it and point on a day-to-day basis. Pointing a bit at a time using a trowel means mixing many different batches of mortar, each of which can have a color that's a little off from the others. These inconsistencies detract from the ultimate appearance of the stonework.

Pointing with a pastry bag makes the work go much faster. When I use a bag, I mix fewer batches of mortar, so the color of the mortar remains relatively consistent.

Using a pastry bag is cost-effective—You can buy a pastry bag, or grout bag, in most masonry-supply stores and in some building-supply stores. The pastry bag is inexpensive (mine cost about $4), but the best thing about using it to fill in joints is that the bag saves a lot of time.

There are other benefits. When you point at the end of a job, you have the option of picking out the mortar color after you see all the stone laid up. My customers enjoy this option because they can change the color they want right up until the day of pointing. If I'm pointing each day with a trowel, the mortar color for the job is set in stone.

The pastry-bag method is economical. Unlike pointing by hand, using the bag results in little waste and less cleanup. Mortar mix delivered through a bag forms a smooth, hard finish, creating a better seal. This is important if the stone is exposed to weather.

Clean stonework is easier to point—The day before I point, I wash the stonework with water and a hard-bristle brush. I take care to rid the joints of any dirt, loose cement or chinks. I also fill any large voids when the stone is dry.

If the joints run consistently 1 in. or larger, I cut a little off the tip of the bag to allow more mortar to flow out, just as I would with a tube of caulk. I don't use a pastry bag if the average size of the joints is wider than 1½ in.

Mortar should be as thick as pancake batter—Chunks in the mortar can clog the pastry bag, so make sure that the mortar mixer (or the tub and tools if you're mixing by hand) is clean. The sand used should be clean, too. If necessary, I sift sand through wire lath to filter larger pieces.

To make it easier for the mortar to flow through the bag properly, I use a rich mix, usually two parts sand to one part cement. Begin with a good dry mix and add color if desired. Then add water slowly until the mix has the consistency of pancake batter. It's a good idea to use rubber gloves to keep your hands from getting stained with mortar color. Also, the lime in the cement burns.

Load the bag and test the pressure you'll need to apply—To fill the bag, hold it upright over the mortar pan so that any spills fall back into the pan. Use any type of sturdy plastic container to scoop out mortar, and fill the bag a little more than halfway (left photo, facing page). If the mix is wet enough, the mortar should drop slowly out of the bag as you fill it.

After you load the bag, twist the open end of it and hold it with one hand to create adequate pressure. Hold the narrow end of the bag up with the other hand to prevent mortar from flowing out until you're ready. Test the bag by squeezing some mortar back into the mortar pan (right photo, facing page). Get the feel of how the mortar flows and how much pressure to apply.

Use the bag like a big, two-handed caulking gun. Run the tip of the bag along the inside of the joints, keeping constant pressure by continuously twisting the open end of the bag and squeezing it in the middle (photo top right).

Shiny mortar is too wet—Occasionally as you fill the joints, go back to the starting point and touch the mortar lightly with your finger. When it feels firm like putty, it's ready to point.

Using a pointing trowel with a ¼-in. to ½-in. blade is preferable. The larger the average joint, the larger the blade should be. I use the trowel to smooth the mortar to the desired look. I prefer a slightly recessed joint, so while I'm smoothing the mortar, I take the tip of the pointer and scrape it along the inside edge of the stones to reveal their shape better (photo bottom right). If the mortar in the joints starts to shine while you are pointing, you should stop because the mortar is still too wet. Remix the mortar occasionally in the pan while pointing. It also helps to rinse the pastry bag with water before reusing it to prevent any remaining mortar from hardening inside.

When you finish pointing, you can use a brush to sweep the joints clean of loose cement. I use a paintbrush when I don't want to change the look of the joints and a hard-bristle brush when I want a rougher look.

The last step is to brush down the stonework with water. You can do this whenever the mortar is hard, but I usually wait until the next day. Washing the stonework not only cleans excess dust and dirt from the stones but also causes the mortar to cure more slowly. The longer mortar takes to cure, the harder it gets. Stained stone can be cleaned with diluted muriatic acid.

The pastry bag can be used to repoint old brickwork or to grout brick and stone patios. However it is used, the pastry bag will expedite the filling of joints, and the finished product will be one of beauty, consistency and durability. □

Christopher Kachur is a stonemason and freelance writer in Newtown, Connecticut. Photos by Kevin Ireton.

Think of the bag as a two-handed caulking gun. With one hand holding the wide end of the bag tightly closed, apply pressure on the mortar and guide the tip with the other hand. Run the tip of the bag along the inside of the joints. Be sure to keep twisting the large end of the bag with one hand and squeezing it in the middle with your other hand.

Smooth the mortar to get the look you want. After filling the joint with mortar, smooth over it with a pointing trowel to get the joint to the desired smoothness. For a slightly recessed joint that reveals the shape of the stones better, scrape the tip of the pointer along the inside edge of the stones.

Laying up Stone Veneer

A masonry facade is an economical way to give wood-frame walls the look of solid stone

by Steven Snyder

Salvaged stone makes a new house look old. Patterned after a 17th-century house in Bucks County, Pennsylvania, this Mississippi home sports 45 tons of stone facade. Because freshly quarried stone would have looked too new, Pennsylvania sandstone was salvaged from an old barn, a dismantled prison, stone quarries and fields.

On a trip through Bucks County, Pennsylvania, in the early 1980s, Brenda and Buddy Williams noticed a stone house that soon became something of an obsession. During subsequent visits over the next several years, they studied the house, which is called Burgess Lea. Built in 1689, the house is an exquisite example of early Quaker architecture, and it had been brought back to museumlike quality by its owner.

I had become the resident stonemason at Burgess Lea, working there every year on jobs that took from a few weeks to a month or two. In 1989, the Williamses approached me with their plans to build a house in Mississippi patterned after Burgess Lea. They asked for my help in understanding the stone details of the house and how they could reproduce them economically. Fortunately, I had plenty of experience in reproducing period stonework in 6-in. and 8-in. veneers. Because of the huge economy of labor and materials in veneer construction, not to mention the insulating advantages of frame construction, the Williamses decided that veneer construction was the most reasonable way to go. An 8-in. veneer on a 2x6 frame could approach the 18-in. thick stone walls of Burgess Lea.

After showing me their blueprints and thanking me for my input, they prepared to leave. On her way out, Brenda turned to me and asked if I would consider doing the stonework. I thought she was joking, but two years later, I was on my way to Mississippi.

Salvaged stone doesn't go as far—The sandstone of the Williamses' house was to match that of Burgess Lea's. Because freshly quarried stone would not have matched the stones used on Burgess Lea, we used salvaged stone from Pennsylvania. As I collected the stone I'd need for the job, I considered its color, size and shape. An old barn, a dismantled prison, stone quarries and fields all yielded parts of the new house.

A ton of stone will cover 25 sq. ft. to 30 sq. ft. in an 8-in. veneer. Most quarries that sell building stone cut or split the pieces into 4-in., 6-in. or 8-in. widths for veneers. These stones don't produce much waste, so 30 sq. ft. per ton should be adequate for an 8-in. veneer. But because salvaged stone must be trimmed to width, it produces more waste, and I counted on covering only 25 sq. ft. per ton. My job was to veneer the facade of the main house (photo, facing page), the center addition and the foundation around the kitchen wing. The rest of the siding would be wood. With a little more than 1,100 sq. ft. of stonework, I needed approximately 45 tons of stone.

Some of the stone details could be reproduced using templates. Working with my longtime friend and helper, Tom Ashburn, I copied the arch stones and the shape of the date stone from Burgess Lea, then produced most of these pieces at my yard in Point Pleasant, Pa. Then the stone, wrapped in stretch film on pallets and loaded onto two tractor-trailers, was off to Mississippi.

At the building site, the first job was to make sure we'd have a dry place to work. Mississippi had been getting heavy rain for several weeks by the time we arrived in January, so I tented the

Make a tent to stay dry. A simple 2x4 frame covered with a tarp allows work to move ahead, even during rainy weather. Line run through pulleys at the eaves of the house make raising and lowering the tarp simple.

Setting reference lines. The author's system for ensuring flat, plumb walls starts with reference lines hung at the top two corners of the wall. The trim piece just below the soffit is installed so that its inside edge is 8 in. from the sheathing. A plumb line is dropped to the foundation from this point to mark the corner.

The horizontal string moves up and down. The author set two vertical stringlines at each corner. Between the inside lines he stretched a horizontal line that could be moved up and down on the vertical lines and that established the stone wall's plane. Galvanized metal ties then are nailed into studs.

Stone is heavy, so plan foundation accordingly. An extra-wide footing supports an extra-wide foundation wall that will carry both the framed wall and the stone veneer. The 8-in. wide shelf is capped with solid concrete block 4 in. thick, providing a strong surface for the start of the stone.

Getting windows away from the wall. Windows would be recessed too deeply if they were installed directly in the wood-frame wall. Instead, carpenters built extra boxes out of 2x material, nailed the boxes into the rough-frame openings and then set the windows outside the boxes. Reference lines ensured a uniform setback for windows in the finished stone-veneer wall.

work area. Using 2x4 rafters attached to a plate at the eaves, I enclosed the front of the main section of the house, running the 2x4s about 15 ft. out from the house (top photo, p. 37). The ends of the rafters rested on a temporary stud wall built near the house. Three pulleys along the top of this open roof made retracting a tarp simple. Even during heavy rains, work proceeded comfortably. The tent was more than worth the effort.

Accurate reference lines mean a straight wall—With a dry work area assured, the next step was to establish accurate reference lines to use as we laid up the stone veneer on the main part of the house. It's easier and more accurate to establish the plane of the wall independently rather than take measurements from the sheathing. Even a slightly racked or out-of-plumb frame can cause some awkward problems for the stonemason later on.

I began setting these reference lines at the soffit. The innermost trim board should be hung so that the distance between its inside edge and the sheathed wall equals the thickness of the veneer, in this case 8 in. After establishing this point at each corner of the house, we pulled a line between the two points so that the trim would create a true line across the front of the house. Carpenters then hung the soffit and attached the trim along its lower edge. On the back edge of this trim piece at both upper corners (8 in. from the face of the sheathing), we set finish nails. From those points we dropped plumb lines to the foundation (bottom left photo, p. 37) and set masonry nails to receive the lines. (Experience suggests this not be done on a windy day!)

We pulled two separate lengths of braided nylon mason's line taut between top and bottom reference points at each corner. These two sets of lines marked the vertical face of the wall at each corner of the house. Next, we ran two horizontal lines across the face of the house. These horizontal lines were pulled fairly tight and were tied to one of the vertical lines at each corner (bottom right photo, p. 37). The horizontal lines could slide up and down on the vertical lines, providing movable reference lines. The plane defined by these lines as they are moved up and down is truly flat and plumb. The remaining vertical line at each corner, to which horizontal lines had *not* been attached, established a straight, plumb reference mark.

Setting windows and installing wall ties—Windows were to be recessed 1½ in. from the face of the finished wall. Carpenters used 2xs to box out the rough openings in the frame wall, then hung the finished windows in the openings at the correct setback from the reference lines (bottom photo, left). Because the windows were recessed, the horizontal reference lines could pass freely in front of them. Setting windows this way ensures that all windows will have the same setback from the finished stonework. Plywood protected the windows during construction.

With windows set, we were ready to prepare the surface against which the stonework would be laid. The 2x6 frame wall had been sheathed in ½-in exterior plywood, and we covered that with

15-lb. roofing felt to provide a moisture barrier. We started at the bottom and moved up, overlapping each course of felt by 4 in. to 6 in.

Next came the anchoring system for the stone. We located the studs at the bottom of the wall and, using a 4-ft. level, drew plumb lines on the roofing felt up to the soffit, marking each stud. Heavy-gauge galvanized wall ties were then anchored to the frame using 2-in. galvanized roofing nails. The wall ties should be applied 16 in. o. c. both vertically and horizontally. Most of the wall ties are hung prior to laying the stone, and the ties generally fit between courses of stone. But because corner stones tend to be bigger and less regular, I set wall ties there as I went along to make sure the wall ties fell in corner joints.

Stone-veneer walls need beefy foundations—The foundation was complete by the time we arrived, but it's important to note that stone-veneer walls require a lot of support. In normal-frame construction, the footing is several inches wider than the foundation wall that will carry the weight of the house. For a house that is to be veneered in stone, the foundation needs to be wide enough to carry the frame and the stone veneer.

Ideally, a 16-in. wide foundation wall on a 2-ft. wide footing is built up to within 6 in. of final grade. At that point, the wall would be stepped back 8 in. to create a shelf to carry the weight of the wall (top photo, facing page).

Disguising veneer corner returns—One of the main objectives on this project was to create the look of solid-stone construction with veneer. But when a wall has been veneered, it is common to see a consistent 4-in., 6-in. or 8-in. return at the corners running straight up the building. Corners in a solid-stone house would be less uniform. Disguising this return would help us achieve the look we wanted.

At the southwest corner of the house, the stone wall would join wood siding. I wanted the corner board and the siding to come out nearly to the outside corner of the stone, thus covering most of the 8-in. thickness of the veneer. To accomplish this, we extended the sidewall at the corner (drawing above). After beveling one edge 60°, we lag bolted a 2x8 to a double 2x4 nailer set on the front wall, extending the sidewall to within 1½ in. of the face of the stone veneer.

The outside edges of the corner stones were chiseled back 30° and laid to the 2x8. The result is a 1½-in. reveal at the corner. The joint between stone and wood can be caulked.

Mortar should have lots of body—With our lines laid out and the surface prepared, we were ready to begin laying stone. Whether building a solid-stone wall or applying a stone veneer, I prefer a sticky mortar with a lot of body. I see mortar not as a bonding agent but as a stable fill that accommodates the irregularities of the stone. The old adage is, "Mortar doesn't hold stones together; it holds them apart." Good stone-laying techniques, not strong cement mortar, will result in a solid, long-lasting stone wall.

Lime adds body to mortar. My mix includes

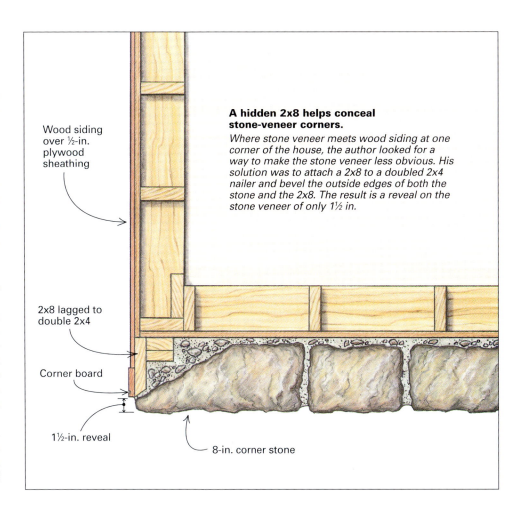

Wood siding over ½-in. plywood sheathing

2x8 lagged to double 2x4

Corner board

1½-in. reveal

8-in. corner stone

A hidden 2x8 helps conceal stone-veneer corners.
Where stone veneer meets wood siding at one corner of the house, the author looked for a way to make the stone veneer less obvious. His solution was to attach a 2x8 to a doubled 2x4 nailer and bevel the outside edges of both the stone and the 2x8. The result is a reveal on the stone veneer of only 1½ in.

Venting the crawlspace. To provide ventilation for the crawlspace beneath the house, wooden vent covers are set on sloped stone sills and set into the stone-veneer wall. Galvanized wall ties connect vent frames to stone, and copper flashing covers the tops of the frames.

Keystone lintels over windows. Wedge-shaped stones, patterned after those on a house in Pennsylvania, provide support over windows. Although flat rather than curved, the wedge shape prevents the stones from sagging into the window frame once the lintel is complete.

A project for nights away from home.
Faced with long stretches of time far from home, the author detailed this date stone by hand in the evenings. The date stone was set in a niche in the middle of the front wall of the house near the eaves.

Pointing keeps water out. Mortar is scraped out at the end of each day to a depth of 1 in. After the mortar has set completely, the entire job is pointed. Because of the mix used and the cold joint between the two layers of mortar, the building will not be difficult to repoint.

one part portland cement, two parts hydrated lime and six to nine parts sand. For mortar on this project, we used a sand with a small-pebble aggregate commonly known as concrete sand. The coarse sand combined with the high lime content to provide the body needed to support the stone as it was placed.

Starting walls with basement vents—The louvered basement vents on the main section were set first. Each vent has a 2-in. by 6-in. frame that needed a stone sill. The two-piece sills were cut and laid up to the proper height; then, each wood frame was set and leveled on the sill (left photo, p. 39). The sills are sloped slightly away from the house to shed water. Wall ties anchor the frames in the stonework. The tops of the vent frames were flashed with copper. Once the frames were set, we laid stones up to the top of the frames. We built keystone lintels over the flat tops of the vent frames, then turned to the corner stones on the main wall.

The corner stones had been cut to rough dimensions in a quarry in Pennsylvania. They now needed to be hand-dressed and fitted into position. The common stonework in the house was roughly dressed (putting a flat face on stone) with a stone or brick hammer to match the stones on Burgess Lea, but the corner stones required a large face that needed to be brought into a single plane. This was accomplished with a 3-lb. hammer, a heavy point and a chisel. The 8-in. thick stones were 28 in. to 36 in. long and up to 14 in. high.

Laying wall stones up to reference lines—After putting down a bed of mortar, we set the stones into it. As each stone was adjusted into position, I sighted between the two lines running horizontally across the wall. Because I could move the two horizontal reference lines up and down, I got an accurate sense of where the plane of the wall should fall. The farthest protrusion of stone came to within ⅛ in. of the line.

I eased each stone down in the mortar bed until it touched the stone below it. With the mortar filling any voids, the stone is not likely to settle further as weight is added to it. All voids behind the stones should be filled with spalls of stone and mortar to prevent settling. In this manner, the amount of stonework laid in a day is limited not by the curing of the mortar but only by the stonemason's energy and ability.

At the end of each day, excess mortar was raked from between the stones to a depth of 1 in. By that time, the mortar had begun to set up and it fell away cleanly from the stonework. These 1-in. deep joints would be filled later when we pointed the walls.

Plan for windows—It's important to plan for windows, doors, rooflines and other architectural features so that the stone veneer will achieve a uniform flow. For example, it's distracting to see all 10-in. high corners and then a 4-in. high piece that is used to hit the top of a window or the top of a roofline. Over windows, stones should be used to break the vertical lines that the window frames create.

The tops of the window frames were flashed with copper in preparation for the lintels. On this project, keystone lintels were used (right photo, p. 39), eliminating the need for angle iron. However, if angle iron is used, I suggest that it be recessed at least 1 in. behind the front of the window frame and pointed over. Even painted iron will rust, and it's better off hidden from view.

As we headed up between the upper windows, we made a niche for the date stone. Date stones are traditional on period stone houses in Pennsylvania. Learning to carve them has been a natural extension of my work and has proved to be valuable. In fact, I had carved the date stone for Burgess Lea in 1985. Because I had a lot of free time in the evenings in Mississippi, I carved this date stone while the house was being built (left photo, this page). The design of the date stone originated in Pennsylvania. It means good luck, and the symbol appears throughout the Williamses' house. Because the date stone wasn't finished when we were ready for it, we built a plywood dummy and set it temporarily in the niche we created to hold a place for the stone. After the arch stones were set, the dummy was removed, and the real date stone was inserted.

The stonework was then brought up to the roofline. We tucked the top stones behind the trim and forced mortar up on top of the last course of stone and down between the stone and the trim. This created a solid top course and provided a backing for pointing.

Point stone after mortar is well-set—Once the stone had been laid up to the roofline and was washed down with water, we were ready for pointing. (Had there been a heavy mortar buildup on the stonework, I would have washed it down with a muriatic-acid mixture.) Pointing weatherproofs and protects the less-stable building mortar within (right photo, this page).

It's important to stress that building and pointing are two separate processes. Pointing is much like adding a final coat of stucco to the brown coat. The building mortar should be well-set before it is pointed. Years down the road, stonework will need to be repointed, and if the process was handled correctly at the start, it should be simple to remove the original pointing without damaging the stonework or the underlying mortar (for more on repointing, see pp. 114-117). I have had the unpleasant experience of trying to chop out mortar in work where the mason pointed at the end of each day's work. Because of the liberal use of portland cement in most masonry work today, difficult repointing is not an experience I wish to pass on to future stonemasons.

I used a rich lime/sand mixture, which resulted in a bright white, slightly raised pointing. The mixture consisted of six parts of fine sand, one part white portland cement and two parts lime. You must keep the proportions consistent to maintain uniform color in the pointing. The mortar is mixed to the consistency of soft cream cheese and applied with a ¾-in. wide trowel. □

Steven Snyder is a stonemason and sculptor in Point Pleasant, Pa. Photos by the author.

Installing a Hammer-Cut Stone Floor

Basic tools produce a quartzite surface that's easy on the toes

by Paul Holloway

As masonry contractors, my partner Ed Brundle and I build everything from concrete-block K-Marts to residential stone floors. The former is our bread and butter, but it's mundane work, the product of which is upstaged by the latest in underwear and toasters. Stone masonry is far more gratifying. Unfortunately, it tends to be like chrome wheels and a new paint job—when times get tough, people do without it.

When times got tough last winter, though, I enlisted the help of my partner and a hod carrier (laborer) and installed a stone floor over the plywood subfloor in my own living room (photo right). My wife and I picked stone so the floor would be as indestructible and easy to maintain as it would be beautiful, important factors when you have three small children.

Our floor would be hand cut using hammers and chisels, just like the stone floors we put down for our customers. It's quicker and easier to cut a floor with a wet-cutting, diamond-bladed saw, but I don't like the look of saw-cut floors—they're blocky and the joints are too uniform for a natural material such as stone. Besides, I couldn't afford to buy diamond sawblades for nonpaying work.

Rocky Mountain quartzite—A stone floor should fulfill at least two requirements: it should be attractive, and it should be flat enough that people can walk barefoot in the dark on it without stubbing their toes.

For our floor, we picked a locally available metamorphic stone called Rocky Mountain quartzite. Quarried in Idaho, it's beautifully colored, reliably flat and tends to break at 90° angles to its face, making it relatively easy to shape with a hammer. Also, the natural texture of the stone makes an excellent non-skid surface.

The stone measures from ½ in. to ¾ in. thick and is available with either silver- or gold-colored streaks in it. The silver variety is consistently flatter and smoother than the gold. That makes it easier to work with and produces a smaller percentage of waste. Nevertheless, like many of my clients, we opted for gold because we prefer the way it looks.

Whatever its color, Rocky Mountain quartzite sells either by the "ton" or as "select." Around here, tonnage costs $.14 per lb. and is based on the luck of the draw; once you buy

The author's new living-room floor is a mosaic of gold Rocky Mountain quartzite, hand cut using hammers and chisels. The floor's even surface allowed baseboards to be installed without scribing.

the stone you're stuck with it whether or not it's usable (stone is packed tight on pallets, so it can be tough to tell). Select quartzite costs $.28 per lb. You sort it out yourself from the "select" pile at the yard. For our 300-sq. ft. floor, I bought 4,700 lb. of tonnage and 1,200 lb. of select stone so I wouldn't have too many rejects to dispose of once the job was completed. Total waste on the job was 2,600 lb., all of it tonnage.

Preparation and layout—A stone floor can be installed over a concrete slab or a wood subfloor. In either case, the stone is set in a ¾-in. to 1-in. thick bed of mortar so that the total thickness of the mortar bed and stone is 1½ in. Over subfloors, though, the mortar is reinforced and anchored to the floor with gal-

Tools of the trade. **The floor was installed without the use of power tools. Pictured here is the author's toolbox with a pair of levels stowed in the lid. In the foreground, left to right, are a pair of stone hammers, a brick hammer, two flat chisels, a brick set and a tuck pointer on top of a brick trowel.**

Photos this page: Roger Schediwy

vanized diamond metal lath nailed to the floor with galvanized 8d nails (drawing right). Over slabs, no lath is required.

Before we installed the metal lath, we snapped chalklines on the walls 1½ in. above the subfloor to correspond to the finished elevation of the stone. This chalkline and the use of a 4-ft. level would ensure a level floor.

Composing the floor—Sizing stone is an art that calls for a good eye and a sense of balance. There's no formula for it that I know of, but generally speaking, the larger the room, the larger the minimum size stone we use. For setting stone, we use a simple rule: the *leave* is worth twice the fit. Or, a well-placed stone not only fits, but its shape accommodates the next stone without forcing the use of odd-shaped stones or stones smaller than about 8 in. by 8 in. (called *chinks*). A floor full of chinks is a sign of sloppy workmanship.

I wanted my floor to be composed with the color randomly distributed and with few chinks (ordinarily, I wouldn't use *any*, but that would have called for sorting through about 25% more stone, which was beyond our budget). I also planned to place unusually large stones at the entrances into the living room. These would serve to invite people into the room.

Working the stone—My stone floor was cut and laid in sections measuring about 6 ft. across, which was about as much stone as we felt like cutting at once. While laying the floor, we checked each stone for flatness, color and structural integrity. Stones with surface irregularities were discarded, except for the few that were especially nice. These were rescued by facing them (flattening their surfaces) using a brick or stone hammer and a combination of flat chisels (bottom photo, previous page). To check for fractures, we visually inspected each stone, then used the old method of striking each one with a hammer; a dull thud instead of a ringing sound signified a reject.

Wherever possible, we used *naturals* (stones that required no cutting to fit). When cutting was required, we set the stone over its final destination and marked the cut lines with a pencil. Then we cut the stone using a brick hammer in combination with an inverted length of 3½-in. angle iron (top right drawing).

Our cutting procedure is pretty straightforward, but mastering it takes practice. The idea is to chip away the stone gradually rather than break it off all at once (large cuts can sometimes be accomplished with a single blow, but they're risky). So we slowly slide the stone forward over the angle iron while striking the stone with the brick hammer directly over the contact point. We reduce most large stones to smaller ones by simply standing them on edge and striking them with the side of the hammer. Difficult stones are laid down and broken with a hammer and a brick set.

Mixing the mud—Throughout the job, our trusty hod carrier kept Ed and myself supplied with stone and mortar. The mortar was mixed

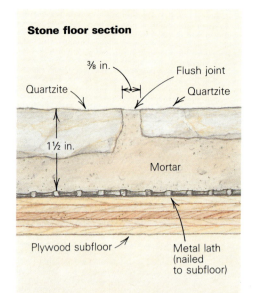

Stone floor section

⅜ in.

Flush joint

Quartzite

Quartzite

1½ in.

Mortar

Plywood subfloor

Metal lath (nailed to subfloor)

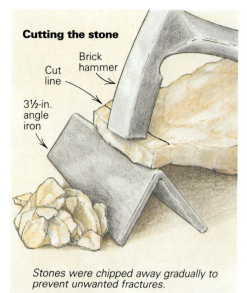

Cutting the stone

Cut line

Brick hammer

3½-in. angle iron

Stones were chipped away gradually to prevent unwanted fractures.

Dry-fitting the floor. **After metal lath was nailed to the subfloor, chalklines corresponding to the floor's finished elevation were snapped on the walls. A section of the floor was then cut when necessary, using hammers and chisels, and laid dry.**

Bedding the stone. **Section by section, the stone was bedded in mortar, then leveled to the chalklines. Poorly bedded stones were pulled up and rebedded as needed.**

Drawings: Bob Goodfellow

with a paddle mixer, which looks sort of like a riverboat paddle suspended over a barrel that's split in half lengthwise.

Our mud recipe was 3 cu. ft. of sand and ¼ cu. ft. of lime per 1-cu. ft. bag of masonry cement. In high traffic areas, we reduced the sand content to 2¼ cu. ft. per bag of cement to make the mortar more durable. The stiffness of the mix was critical. It had to support the stone, but it couldn't be so soupy as to be self-leveling nor so stiff that it wouldn't ooze up between the stones during installation.

Each batch was workable for about an hour and a half to two hours. To maintain the desired consistency, we tempered the mortar as often as necessary (based on temperature and humidity) by working it with a trowel or shovel and sprinkling a little water on it.

Putting the stones to bed—Sections of the floor were laid dry (top photo, facing page), then bedded in mortar one stone at a time. We started by bedding several stones adjacent to a wall, leveling their tops to the chalkline on the wall. Then we bedded several stones farther out from the wall, each within 4 ft. of a level stone. Once these stones were leveled, we bedded the remaining stones in the section using our 4-ft. level as a straightedge. The joints between the stones typically measure ⅜ in.

To set a stone, we started by throwing a trowel of mortar on the floor and thoroughly working the mortar into the metal lath using a brick trowel. Enough mortar was placed to produce good *squeeze*—that is, plenty of mortar squeezing out the sides of the stone. Next, we dropped the stone on the mortar and either leaned or stood on it (depending on the size of the stone) to lower it close to its finished elevation. Finally, we leveled the stone by tapping on the level or on the stone itself with the butt of the trowel handle (bottom photo, facing page).

It's critical that a stone floor have absolutely no voids in its mortar bed. Accomplishing this is a challenge, but it's essential for a durable floor. We checked for voids by tapping on each freshly bedded stone with a hammer. A hollow sound meant the stone needed to be lifted and the bed reloaded with mortar.

The manicure—Our technique for finishing the joints between stones, which we call the manicure, is similar to that of a tilesetter. First, we completely pack the joints with mortar (the squeeze takes care of most of that; voids are filled using a tuck pointer). Then, once a section of the floor is bedded, we remove excessive mortar and smear mortar over the entire surface of the stone using a damp sponge (photo below left). We allow the mortar to harden for a couple of hours, then mop the surface of the floor with a wet sponge, taking care not to recess the joint (a flush joint is what we're after).

At this point, we typically let the floor sit overnight, and then have our hod carriers scrub it down thoroughly the next morning using stainless-steel pads or those green "scrubbies" and about a 9% to 12% solution of muriatic acid (ten to eight parts water to one part acid, depending on how stubborn the stain is). This requires that the floor be wet with water first and kept wet to prevent the muriatic acid from soaking into the floor, which would burn the mortar and stain the stone. Finally, the muriatic acid is mopped up with sponges and the floor scrubbed with water until its surface is impeccably clean. During this phase of the job, we have extra sponges and buckets on hand to soak up water and prevent damage to the adjacent drywall.

Anyone who has worked with muriatic acid knows that even though it won't burn holes in your clothes, it can still be nasty stuff. You need to ventilate the work area and wear rubber gloves and eye protection when working with it. Because this floor was in my house, I managed to avoid using muriatic acid (except on a few stubborn spots) by scrubbing the floor thoroughly with abrasive pads and water in the evenings instead of letting the floor sit overnight. The results (photo below right) were comparable to an acid-cleaned floor.

Finishing up—A completed floor can take from a few days to a month to dry, depending on temperature and humidity. This one took several days. Stone floors, like brick ones, effloresce while drying; that is, moisture evaporating from the surface leaves behind a white, powdery deposit of calcium carbide that comes from the mortar. Once my floor was completely dry (when dry, mortar gets lighter in color), I sponged off the efflorescence with vinegar. Then I allowed the floor to dry once more before applying a sealer.

Ideally, a sealer should waterproof the floor while highlighting the color of the stone. We use a product called Glaze 'N Seal Concrete and Masonry Lacquer (Glessner Corporation, P. O. Box 6427, Moraga, Calif. 94570; 415-621-1414). I applied it to my floor with a paint roller. A coat of this sealer once a year will keep the floor looking as good as new.

My floor took 80 hours to complete. This included installing the baseboard which, true to form, lay flat against the floor without scribing. How's the floor holding up so far? Carson City recently had a series of earthquakes measuring up to 4.8 on the Richter Scale. Not a single crack appeared in the floor.   □

Paul Holloway is co-owner of Northeast Masonry in Carson City, Nevada. Photos by author except where noted.

Finishing off. Once a section was fully bedded, mortar was floated over it with a damp sponge to top off the joints.

A few hours later, the floor was scrubbed clean with abrasive pads and water. When dry, it was coated with a waterproofing sealer.

Laying Flagstone Walks
Tips from a professional mason

by Joe Kenlan

Flagstone walks enhance the appearance of a yard or an entry. Above, a pathway with stones in place but not grouted; above left, the finished walk. A dry-laid path (above right) is less formal and rougher to walk on than a mortared walkway.

Flagstone is not a particular kind of stone, but a generic term. It designates any hard stone that fractures into broad, flat pieces suitable for paving walks, patios and floors—most commonly slate, limestone or sandstone (marble is often used for expensive installations). Most large masonry-supply stores carry flagging material; the particular type will depend largely on what is available locally. Sandstone and limestone can be found throughout the country, but the metamorphics—such as marble and slate—are limited for the most part to mountainous areas. Shipping stone is expensive, and if you order it from far away you can expect the cost of shipping to exceed the cost of the stone itself. For example, I can buy Tennessee crab orchard stone for $50 to $60 a ton where it is quarried. But it sells for $150 to $165 a ton here in North Carolina, just 300 miles away.

The most important characteristic of good flagstone is hardness. If you can crumble the edges of the stone with your hands it definitely won't stand up to traffic. Good flagstone also has to be flat and smooth so that tripping will not be a problem.

Flagstone can be laid over a bed of either concrete, sand or gravel, or directly in the earth. It lends itself to everything from formal entrances to meandering garden paths. Laying a stone walkway requires little expertise (skill will come with practice) and no exotic equipment. You'll need a rule and level to lay out and set the screeds (the boards that define the contours and level of the walk), a circular saw for cutting screeds and stakes, and a trowel and a brick hammer for setting and trimming the stones. Concrete can be mixed in a wheelbarrow with a hoe. An ordinary steel garden rake is handy for spreading the bed and for pulling the reinforcing wire up into the concrete. Another useful item is a pair of cutting pliers for trimming the wire. Grouting requires a stiff brush, large sponges, buckets and a piece of burlap for final cleaning.

Laying out a mortared walk— Most flagstone walks are functional rather than decorative, and should go from point A to point B as directly as possible because pedestrians are likely to do that whether you put the walkway there or not. Moderate curves for appearance are okay, but any large deviations should be justified by taking the walk around something, such as a tree, a bush or a boulder.

The walk should be wide enough for its expected traffic. A path around a little-used part of a yard might be only 16 in. wide, but a main entry requires a minimum width of 32 in. The busier the walk, the wider it should be.

Once the layout is decided on, mark it out on the ground using string for the straight sections and marking powder (mortar, flour or lime all work well) for the curves. Alternatively, you can cut its outline into the ground with a shovel. The layout width includes the finished width of the walk plus a 3-in. allowance for screed boards.

Digging the trench—Once the walk has been marked out, dig a 5-in. to 6-in. deep trench. This depth allows for a 3-in. or 4-in. reinforced concrete slab and the inch or two the stone top requires. Cut the bottom of the trench flat (a

Photos: Michael Yarborough

square shovel is useful here) and square up the sides. Unless you are working in solidly compacted earth (for example, an old existing walkway), thoroughly tamp the bottom of the trench by hand or with a power tamper.

In wet or filled areas or when working on uncertain subsoils, my helpers and I usually add at least 2 in. of crushed stone to the bed of the walkway to provide a firm base and to allow some drainage. If crushed stone is used, the trench must be deepened accordingly. Some areas may require more extensive measures, such as culverts, to provide drainage.

Next the screeds are set. These boards act as forms for the concrete pour, defining the shape and the finished height of the walk. We use 2x6 lumber for the straight sections because it is stiff enough to resist bowing under the pressure of the poured concrete. We use strips of plywood kerfed vertically for the curved sections (inset photo at right). It's faster and easier to kerf an entire sheet of plywood and then rip it into strips for curved screeds.

Unlike brick walks, stone walks cannot easily be crowned since the stones don't follow a regular pattern across the walkway. We usually try to set the screed boards even with or just above the adjoining grade with a slight slope (about ¼ in. per foot) to one side, so water will drain. This keeps the walk dry but doesn't interfere with lawnmowers or with the occasional traffic across the walk.

We prefer to work with stone that is between 1 in. and 1½ in. thick for mortared work. Anything thinner than ¾ in. may be knocked loose. Thicker stone costs more (stone is sold by weight) but is perfectly suitable otherwise. Using thicker stones here and there in a walkway makes screeding more of a problem because the mortar bed must be adjusted to accommodate the additional thickness.

Sometimes there will already be a set of steps at one end of the walk. In laying out the walk, remember that steps at either end of a walkway will determine its height since the surface of the walk becomes, in effect, another tread in the step and must be set the same distance down (or up) as the rest of the risers. Add another step if you have to to even things out. Since steps are virtually always set level side to side, set the walkway level at this point, and begin to slope it gradually about one tread width from the first step.

Stakes, either of wood or rebar, are used as needed to hold the screed boards in position, and earth is backfilled against the outsides of the boards to help hold them and resist bowing. To set the boards at the proper height, we either tack them to the stakes or tie them to the rebar in such a way that they can be easily removed later. They can be shimmed up using small stone wedges.

The screed boards are checked with a level to ensure sufficient drainage and with a rule to make sure the two sides are equidistant. This is particularly important when laying out curved sections. We lay out one side completely and then use the rule and level to set the other.

Once the screed boards are set and checked, we cover the bottom of the trench with polyeth-

After the walkway has been laid out and excavated, it is edged with screed boards to contain the pour and mark the finished level of the stones. Straight sections are done with 2x6s; curves (inset) are edged with strips of plywood kerfed vertically. Rebar braces and a polyethylene moisture barrier complete the preparations. Above, concrete is placed over the reinforcing wire. On this walkway, flagstones are being set directly into the wet concrete.

ylene. This limits moisture passage up into the walkway, which can cause unsightly efflorescence in the flagstone.

Pouring the concrete—On top of the plastic we roll out standard 4-in. by 4-in. slab reinforcing wire (or "hog-wire" fencing), cut it and fit it.

Now we go one of two ways. If the project is large we generally pour the whole walk in one pass from a truck, let it set up and then place the stone in a mortar bed on top. If the project is small enough to do in one day, we pour it and set the stone in one operation (photo above). The concrete mix (1 part cement to 3 parts sand, or 1 part cement to 2 parts sand to 3 parts gravel) should be fairly stiff, as for normal slab work.

Using the first method we essentially pour a concrete sidewalk, except that instead of

screeding off at the top of the boards as would normally be done, we use a notched 2x4 to screed off so that the finished top of the pour is 2 in. below the forms.

To make a screed guide we take a 2x4 that is 6 in. or so wider than the walk and screed boards and notch it 2 in. at both ends. The notched-out part of the board should be an inch or so narrower than the finished walk. When this board is pulled along the tops of the screed boards it smoothes off the concrete 2 in. below the finish height of the walk.

Next we pour the concrete to the approximate depth, pulling the reinforcing wire up into the concrete as we go, and screed it off. The surface is left rough so that the mortar bed for the stone will bond to it. As we pour, we remove any stakes that are inside the screed boards. The concrete should hold the screed boards in place at this point. If the screed boards are bulging anywhere, we add stakes as necessary to hold back the pressure of the pour.

Setting the stones—Once the slab has hardened (a day is usually sufficient), the stones can be set in a shallow mortar bed. Before starting to set them, it's helpful to look at the type of stone you'll be using and to decide if you want to lay it in a particular pattern. In our work, we usually let the stones decide the finished pattern. What we do emphasize is the relationship of the stones to each other so that we have evenly spaced, parallel joints. Using a brick hammer, we shape and fit each stone so it interlocks with its neighbors (sidebar, facing page). There is no structural reason for this, so many masons lay the stone randomly to save time. But we feel the result is well worth the extra effort. After all, the walk will be there a long time.

As in most of the other stonework we do, we work from the outside in, laying the two edges and then filling in the space between them. Edge pieces need to be large enough to keep them from being accidentally dislodged over time. In laying out the stone we first take stock of its natural shapes. Some stones break at sharp angles—others more gently—so we try to leave spaces that will conform to the stone we have.

The stones should be well washed before using them to rid them of quarry dirt and dust, which will interfere with the bonding. We like to keep them damp so that they don't suck moisture from the cement while they are being set.

The stone is set using a rich mortar (1 part cement to 2 parts sand) that is mixed to about the consistency of mortar used for laying block. (If you find that the stones ''float'' while you are setting them, use a slightly drier mix.) With this mix, pour a bed thick enough to bring the top of the stone slightly above the screed boards, and

Grouting. After the stones have been allowed to cure for a day, a slightly damp grout mix is worked into the joints with a stiff brush (top left). Then more mortar is worked over the entire surface of the walk with a wet sponge, forcing it into the joints (second from top). Subsequent spongings remove most of the excess mortar (third from top); a final rub with burlap (bottom left) polishes the stone.

begin to set the stone. Using a 2x set across the width of the walk as a guide, gently tap the stones down to finish height using the handle of a hammer or trowel.

As we work along, we also use the board to make sure that all the stones are in the same plane by laying it diagonally across the walk in several directions. This is important to prevent toe-catching irregularities.

Grouting—Once the stone has been set and allowed to cure for a day, we grout the joints with a mix of 1 part sand to 1 part portland cement. The grout is mixed slightly damp and is swept and worked into the joints with a stiff brush (photo facing page, top left). When the joints are well filled, additional mortar is placed on the surface of the walk and is worked into the joints using a large soft sponge. We use a generous amount of water on the sponge to make sure that the grout is wetted all the way down into the joint (photo second from top, facing page).

The sponge is gently worked over the tops of the stones to smooth the joints. The first pass leaves a slurry of cement over the surface of the walk that is gradually removed by subsequent spongings. We use plenty of clean water until all the visible mortar is washed from the surface of the stone and the joints are filled smoothly and uniformly (photo third from top, facing page).

Generally, three or four washings with the sponge and water are needed to clean the walk. One final cleaning, using a piece of damp burlap, polishes the stone and removes any remaining mortar film (bottom photo, facing page). Some masons use straw instead of burlap, but we don't like to use straw because it leaves debris in the joints.

If, after the walk has been cleaned and dried, there is still some film remaining on the surface of the stone, we let the walk set for several days (until the mortar changes color) and wash it down with a non-acid masonry cleaner. We test the wash on a scrap piece of stone to see what effect it will have. The wash and a rinse will take care of any remaining film. Rubber gloves and eye protection are essential.

In exterior work we generally do not apply a sealer, but if one is required we test it on several unlaid stones to see how it affects their color and surface. Sealers can cause some stones to become slippery when wet.

After the walk has been allowed to set thoroughly we remove the screed boards and backfill the edges with dirt.

Dry-laid walks—Stone paths without mortar are usually reserved for less formal entrances than the grouted type since they are less even and smooth (photo at right, p. 44). For dry-laid walks, we select stones at least 2 in. thick since the mass of the stone is what will anchor it in place. The stones can either be set individually with a moderate space between them (this method is useful if the stones vary a great deal in thickness), or by using a modified version of the concrete-set method described above. In the latter case we dig the trench as before, except that the depth is determined by the thickness of

Cutting flagstone

The surface of a stone can be smoothed out with a mallet and chisel.

Flagstones—whether slate, limestone or sandstone—are generally strongly layered, a characteristic that produces flat, thin pieces. But such stones resist attempts to cut them across their natural rifts. Consequently, in our work my helpers and I first look for a stone that nearly fills the space we are dealing with and then trim it by nibbling in from the edges with a brick hammer or similar tool. Attempts to take off large sections at once usually result in unpredictable breaks.

Undercutting the backside of the stone before working the face is helpful, as this provides a weak point to break against. Smoothing can be done by working the stone at an angle from the edge. Don't ever work toward the face of the stone because more often than not, you will break out a larger chip than you intended. For more precise cuts, use a skillsaw equipped with a masonry cutting blade. Score the stone about halfway through and then snap off the waste piece with a sharp hammer blow. Again, always work away from the face of the stone. If you have access to either a stone saw or a splitter you can of course use them. But a sharp brick hammer or hammer and chisel will usually suffice.

When sawing stone it is important to use plenty of water, even when using dry-cutting blades, to reduce dust. Ear protection and a dust mask should be worn, and eye protection is a must at all times.

On most flagstones, surface imperfections can be removed by chipping them off with a sharp thin-bladed chisel held at an angle of about 30° to the face of the stone (photo above). You simply flake off the undesired layer.   —*J. K.*

the stone to be used plus a 2-in. to 3-in. leveling bed of crushed rock and sand.

Next, the trench is tamped and crushed stone is added as required. Then we roll out polyethylene, tar paper or polyester cloth (which inhibits the growth of weeds and keeps the sand leveling bed from washing into the stone below), and lay a bed of sand on top of that. The sand is screeded, and the stones are set using more sand where needed to bring them to ground level. Stones should be laid tightly together to help them bind on each other. Then sand or fine dirt is spread on top and allowed to work down into the joints. We leave the walk covered with sand this way for as long as is practical so as much as possible works down between the stones.

Walks laid this way have a tendency to spread a little over time but we feel that this adds to their simple, rustic character. It does, however, make them less sure footing for spiked heels.

Stepping-stone paths—The most informal type of walk is the stepping-stone path, with the stones spaced just closely enough to accommodate footsteps. For this, thicker stones are preferred, and each stone is set individually. Stepping stones are usually set flush with grade and follow the contours of the ground.

For a typical path, we begin by laying out the entire walkway, testing occasionally to be sure that it will be comfortable to use. We try to emphasize parallel lines between adjacent stones. Then we scribe the outline of each stone on the ground with a trowel, remove the stone, dig out and smooth the bed and reset the stone, making sure that it sits firmly and does not rock. In muddy or soft soils, we excavate deeper and add a 1-in. to 2-in. thick layer of crushed stone beneath each stepping stone to stabilize it.   □

Stonemason Joe Kenlan lives in Pittsboro, N. C.

A Stone Cookstove and Heater
This dual-purpose unit combines masonry, steel and a long smoke path

by Jonathan von Ranson

Since turning homesteader in the woods of western Massachusetts, I have had a few definite surprises. Having a boss from my former newspapering days show up at the door was one. Deciding one morning to go ahead and move into the stone house that took us three-and-a-half years to build was another. The greatest—and in some ways the most pleasant—surprise is the recurring one of operating a stone cookstove. It cooks our meals all year round, and in winter it's our 800-sq. ft. house's main source of heat.

We found out about masonry cookstoves as we were researching what kind of cast-iron one to buy. ("We" consists of Susan and me and our three teenagers, Erik, Kristin and Joel). Our research led us to Albie Barden of the Maine Wood Heat Co. of Norridgewock, Maine (see *FHB* #7, p. 49). He invited us up to see his masonry stoves, and we made the journey to central Maine one February day.

Barden fired up the imposing brick range in his kitchen. The oven climbed to 350°F in about 15 minutes, with only four or five sticks of wood. In his parlor is the heater version from which the cookstove takes much of its inspiration. He lit a fire in that, too. People who've built masonry stoves like to show how well the draft pulls: down and around and through . . . "as long as it goes up at the end," says Barden, watching your face.

Not ones to make hasty decisions, Susan and I were more than halfway home again when we turned around and drove back to buy the Finnish cast-iron parts Barden imports for the stove.

The Finnish connection—In this country, masonry heaters have come to be known as Russian fireplaces, because the first ones built here were put together by Russian émigrés in the style of the brick stoves of their homeland. But in fact, the Finns are in the forefront of masonry-stove technology, and it is they who have developed the special mortars and fittings that allow cast iron to be wedded to masonry so that a wood-fired heater can also be a cookstove.

In all masonry stoves, the smoke path is long. The hot gases travel past virtually every stone, leaving some of their heat in each before going up the chimney. In the stove we built, the smoke makes a 5-ft. loop through the base of the chimney to warm that mass after it's already traveled about 14 ft. in a serpentine 8-in. by 8-in. channel that starts at the firebox (drawing, facing page).

The parts that we bought from Barden included the cooking surface and its lids, the oven and its door, the firebox and ashpit doors mounted on a common frame, a grate with a frame, a slide damper and its frame, and two cleanout doors with frames.

The largest of the castings is the cooking surface. Its rigid frame is meant to sit above the masonry on a bead of mortar, which mates with a raised, ¼-in. bead on the casting. This design makes for a tight seal and guards against the top's warping. There are three lids—two normal-sized ones on the right and a big one—16 in. in dia.—over the firebox on the left. They are cast with deep waffling on their undersides to increase the area that can absorb and transmit heat to the top surface. The lids quickly get hot enough to boil water, and they draw off heat quickly enough so the rest of the stove doesn't overheat (tests show that mortar begins to disintegrate at temperatures above about 400°F).

The oven has shallow flutings that work like the waffle grids on the lids. It's small by American standards (13 in. by 10 in. by 20 in.), but it roasted a 16-lb. turkey last Thanksgiving and easily holds four to six loaves of bread. The oven's heavy, tight-fitting castings show evidence of fine workmanship.

The Finns have two notions about fireboxes that surprise most Americans. The first is that they should be small. Our firebox door is only 8½ in. by 6 in., and the firebox itself is just 21 in. deep. Wood burns more efficiently in a small combustion chamber, and the smaller fire is less likely to overheat the masonry. Besides that, it calls for small logs, so we tend to use less wood than we might in an American-style woodburner.

The second idea, which sounds almost like heresy, is that fireboxes should not be airtight. The Finns feel that woodstove efficiency depends on complete combustion, and on the route taken by the hot gases after they leave the firebox.

The positive-closing slide damper in the chimney flue up near the second floor gets shut tight when the fire has burned down to glowing red coals. This prevents room air from being drawn through the stove and leaving the house, setting up a convection current that would also cool the masonry from inside. With the damper shut, the house is heated all night by the stove top, the stove mass itself,

The stone cookstove designed and built by von Ranson weighs ¾ ton. He carted the stone from a nearby quarry and cut it to size in a sandbox on site. The firebox is small—only 21 in. deep —and accepts just four or five sticks at a time, but it's efficient enough to serve as the house's main heat source.

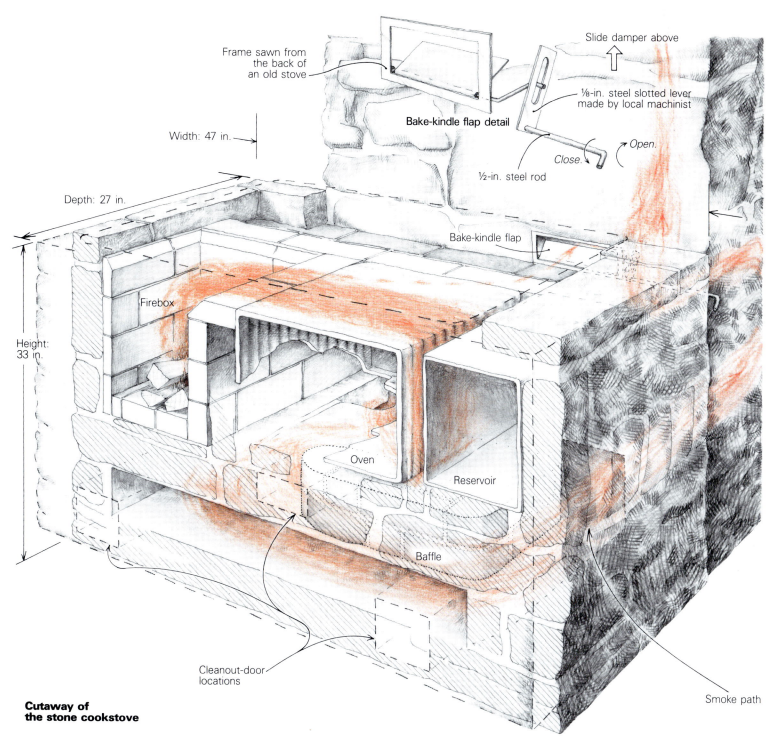

Frame sawn from
the back of
an old stove

Slide damper above

⅛-in. steel slotted lever
made by local machinist

Bake-kindle flap detail

Open.

Close.

½-in. steel rod

Width: 47 in.

Depth: 27 in.

Bake-kindle flap

Firebox

Height:
33 in.

Oven

Reservoir

Baffle

Smoke path

Cleanout-door
locations

**Cutaway of
the stone cookstove**

and the chimney base where the flue loops through, which I think of as our heat sink. They're all still warm to the touch when we come downstairs in the morning.

The ashpit door is beneath the loading door. The grate, with frame, sits between the two.

Both the visible and the hidden cast-iron parts have a spare, Scandinavian look. There was no thermometer in the oven door, so we bored a hole and installed one.

The design of our stove required some additional pieces, such as a small cleanout door that would let us get at the smoke path under the oven. We also got a welder to fabricate an 8-gal. stainless-steel hot-water reservoir with a scoop at the top for filling and a faucet at the bottom for drawing. Last, at Barden's suggestion, I made a flap that would allow the stove to operate in two modes: kindle, and bake-and-heat. The flap is open in the kindle mode, and smoke and heat go straight up the chim-

ney. This sets up a good draft, and gets the fire burning nicely. When the fire is well established, we close the flap, and the hot gases are routed through the entire masonry loop. This is the bake-and-heat mode.

We got the basic components of the bake-kindle flap secondhand at the Bryant Steel Works, a stove graveyard in Thorndike, Maine, and the under-oven cleanout door at the Good Times Stove Company in Goshen, Mass. (they have restored stoves, too).

The parts for the stove came to a little over $1,100, including everything: metal, cement, even travel. Most of this by far was for the Finnish castings. Don't take this route to save money on a cookstove purchase, unless your time counts for zero.

There are a few Finnish tricks we decided not to use. Some Finns sleep on top of their masonry heaters—a bunk is designed right into them. In another plan the smoke goes

through a masonry bench, which acts like a radiator and gives a toasty place to sit. They use flues like plumbing, piping the hot smoke to where it will give useful, even heat.

Another technique—short, hot burns—we don't employ out of protectiveness toward our stove. Theoretically, short burns are more efficient, since the fire burns only long enough to bring up the heat of the masonry. Also, combustion is more complete in a hot fire. But that technique is more appropriate to pure masonry heaters than to our masonry and cast-iron hybrid, in which the iron draws out enough of the heat to eliminate the need for fast, hot burns. Medium-length, medium-heat burns (just right for making pancakes or baking a casserole) seem to work best for us.

Stonework—With the metal parts in hand, it was time for us to decide on the material for the rest. Plans for brick faced with stone

looked cumbersome: too thick and too many mortar joints. All brick, then? No. Our house was made of stone, so we decided on stone for the stove.

Scandinavians almost invariably use brick, often surfaced with tile, for their *Kachelöfen*. For all we knew, we'd be building the only stone kitchen range on earth.

The smoke path is one of the things that makes our stove unique. As shown in the drawing on the previous page, there are three levels of it: under the stove top, under the oven (with a little eddy beneath the reservoir) and—here's the Finnishing touch—down in the stove's basement. On its way across the top level, the smoke heats the stovetop and the upper oven; descending, it heats the water reservoir and the side of the oven; once at the middle level it heats the bottom of the oven; then it's drawn down an opening into the "basement" where it leaves energy for space-heating the house.

I had four years' experience at masonry, but this stove was the most difficult thing I had ever undertaken. Such complex spatial relationships in three dimensions—such variables in mortar joints, expansion joints and missing parts—tax the ordinary mind. (Such heavy slabs of stone tax the ordinary back.) I had hardly anything in the way of plans. Barden had lent us a book, written in Finnish, with two or three plans for masonry cookstoves. From there on I was on my own. Over the next few months, the designs in the book—microscopically small, intended for brick—became clearer. Still, designing in a water reservoir and an optional smoke path produced powerful feelings of insecurity. The best advice available involved using three different mortars—the firebox looked too small to heat anything, let alone the whole house—and as I built, the stone had its own ideas about how things should happen.

I decided to use stones as close to 4½ in. thick as possible; that would be the thickness of the exterior walls and the horizontal divider, or baffles.

We began by hacking our way through a mile of forest to an abandoned quarry, and there I got my initiation in quarry-style stone-cutting. We picked out a 9-in. thick slab, tapped a line along the grain edgewise with a chisel, turning the stone several times, and it dramatically opened like a book, 1½ ft. wide and 7 ft. long. It took five trips with our 1947

The front stones (top left) were cantilevered to bring the heater out to full dimension and to create a kickspace. Combustion gases will be pulled into the crook of the brick baffle, and then directed around it and into the chimney. The frame for one of the two bottom-level cleanouts is in place. On the second level (center), three large slabs span from front to back. They will support the firebox, the oven and the hot-water reservoir. The hole at the back of the middle slab is the smoke passage. The photo at left shows the reservoir and oven in place, with the firebox almost completed. Smoke will curl first over the oven, down between it and the reservoir, then back under the oven before rising through the passage in the middle slab.

jeep to bring back a couple of tons of split stone, a stack maybe 2 ft. high.

I spent the next several weeks in the sandbox (the best place for stonecutting), creating particular shapes, fitting them together dry, and trying to fit them together wet in my mind. Steve Busch, a mason in South Paris, Maine, who had worked with Barden building several Russian fireplaces and was about to undertake a cookstove, suggested keeping the firebox isolated with an expansion space. But I saw no way to avoid cementing the firebox to the face of the stove, to maintain structural integrity and to control the fire. I didn't want cracks through which smoke and sparks could pass and short-circuit the carefully planned heat path. I was in uncharted regions of fire-taming and a little scared.

The ½-in. mortar joint I figured on turned out to be more like ⅞ in., a discovery that sent me back to the sand pile for more cutting, and eliminated almost an entire tier of stone and a couple of days of my life. Handcut stone tends to come out a little bigger than you mark it. Cutting the oven lintel took several tries, but finally one emerged from the sandbox, a virtual stone ruler about 1½ in. by 4½ in. by 17 in.

I laid the stone courses up in stages, allowing the bottom tier to set before going on to the next. First, I set four large stones on the slab for the floor of the stove. Once they set up, I built the walls, the front wall cantilevered forward 3 in.—more because the slab was that much too small than because we needed a place to put our toes. The overhang rested on 2x4s and shingles while the mortar set. A few red bricks were laid as a baffle around which the smoke would circle in the stove's basement (top photo, facing page).

Those brick were mortared with a special imported Finnish cement called Savi Uuni Laasti, which I also bought from Barden at Busch's recommendation, but I used a standard portland-base mortar for the rest of the stove (except for the firebox, which used furnace cement). The Finnish stuff—the mortar of choice for a brick stove—comes in a 75-lb. bag and costs $25. Not only is it expensive, but it isn't recommended for use with stone. So I stuck with a 1:1:6 proportion of portland, lime and fine sand from there on, even in the back where I laid a few more brick. This high-lime mortar is recommended for use in high-heat areas. It isn't quite as strong as ordinary mortar once cured, but it does stand up better to extreme heat. It sounds paradoxical, but it's the seismology principle: the weaker mortar will allow hairline cracks to form in joints without fatal resistance that might take the form of one large stovequake.

Putting things together called for a scalpel-and-tweezers type of masonry, especially at the front. I wanted to offset my vertical joints, but with five openings and several slender vertical stones between them, there wasn't often much bearing surface. Some of the stones didn't get tied in until the very top course, and I had to hold one of them in place with a pipe clamp until the mortar set.

Three horizontal 250-lb. stones capped the

A welder fabricated the 8-gal. hot-water reservoir for the stove. The oven with its door-mounted thermometer is to the reservoir's left, with a cleanout just below it. The small door at the bottom is also a cleanout.

basement smoke channel and spanned the cleanout doors (middle photo, facing page). It took two strong people to set these into place. Then the big day came when the large metal units—reservoir, oven and loading door—were set into place.

The method that day was to work from the right, setting the right wall stone, then the reservoir, then the 4½-in. deep stone between it and the oven, then the oven itself, then the next vertical stone. Doing it this way rather than dropping the metal units into ready-made stone openings allowed for a bed of mortar both under and alongside the metal pieces, sealing out drafts and sealing in smoke. It also reduced the likelihood of jostling the stonework after it had set up. I wrapped each metal unit with fiberglass insulation where it contacted the mortar to create about 1/16 in. of expansion room. Otherwise I didn't see how the stone could stay intact against the greater expansion rate of the metal. This seems to have paid off.

Next day the firebox fell together reasonably well thanks to a day spent earlier cutting the bricks with a carborundum blade in a circular saw. A thin tier of capstones went on the following day (bottom photo, facing page), as did a small section of brick to form the rear wall behind the oven.

The kindle-bake flap was next. It is a regular gravity-held flap for which I designed a lever of ½-in. steel rod that extends out the right hand size of the stove (drawing, p. 49). A twist of the lever and the flap opens and lets smoke go straight up the chimney; a reverse twist and it falls shut, sending the smoke on a 20-ft. journey around the oven before it goes up the chimney, perhaps 200°F cooler.

After everything had set up well, Susan and I laid a bead of cement around the top of the masonry and set the frame of the cooking surface into it, leveling it carefully. We meticulously pointed up the mortar so the iron would be able to expand horizontally without catching on a cement crumb. (All joints, incidentally, were pointed up both inside and out as soon as the mortar had set up.)

That finished the job. We were advised to wait a couple of months before using the stove to give the mortar time to cure.

Afterburn—We kept the first fires small, and felt the stones gradually heating up over a matter of days, not hours. The stove has been in regular use since last fall and has passed its first heating season in fine shape.

The only problems are that the oven gets a little hotter toward the rear, and creosote occasionally drips out of the slide-damper slot in the chimney. If I'd thought of this problem when I was building the chimney, it would have been easy to slope the damper inward a little so the creosote would run back into the

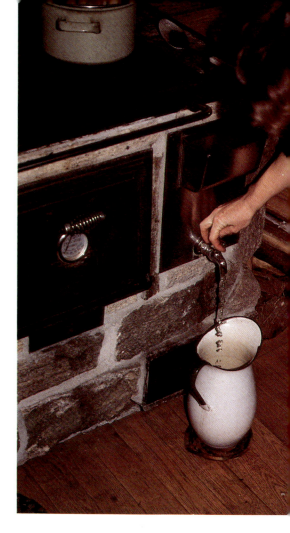

flue instead of out over the stonework. Now all we do is plug the gap. Actually, creosote isn't a big problem. We get none at all if we run the stove fairly hot on kindle mode for 10 minutes or so when we start a fire. We also try to avoid slow, smoldering burns, which let the chimney cool and creosote form.

To correct the uneven oven temperature, it will only be necessary to install a baffle of the kind found in old cookstoves to bring the smoke forward under the oven. This flat metal baffle (maybe ⅛ in. by 2 in. by 12 in.), which stands on edge catty-corner under the oven, would bring the smoke toward the front as it moved from right to left under the oven, (like herding cattle through a gate). Also, either firebrick expands more than I guessed or the iron cooktop didn't "flat" quite right, for a stone lifted slightly in the face above the firebox. This would be easy to repair, but the situation is stable and the stove is solid.

With a year of practice under her belt, Susan now calls the Finnish-style cookstove her favorite part of the house. The hours spent in all stages of anxiety and torment have been repaid—and still the payments continue. We casually stoke the front-loading firebox with the four or five small logs that it will accept at a time and heat the house while making breakfast and dinner. On grey, wintry days a longer fire is required. Our wood consumption is about 3½ cords a year—I like that. □

Homesteader Jonathan von Ranson writes and does stonemasonry in Wendell, Mass. Photos by the author.

Masonry Heater Hybrid

Cross a traditional fireplace with a masonry stove for something home owners can warm up to

by G. Karl Marcus

In the spring of 1985, I was asked to design and build a hybrid masonry heater—part stove, part open-hearth fireplace. This went beyond my experience with traditional masonry stoves, and left me scratching my head for a way to satisfy the design criteria. That's when I met Frank Piwarski. He had recently opened his own design and consulting firm, called Ultra-Fire, and was designing everything from fireplace inserts to commercial waste-wood incinerators. He'd never been involved with masonry stoves before, but Piwarski understood combustion mechanics well enough to design a firebox with a

gasifying combustor and a heat exchanger that gave my client just what he wanted.

During the construction of this hybrid prototype, Brad and Cheri Miller approached me about helping them design and build a similar stove in their owner-built home. They had read about masonry stoves, were impressed with the concept of heat storage and had planned for a fireplace of some sort on their main floor.

The Millers and I discussed the size and shape the structure should take, and we settled on a single glass loading door, a chimney behind the firebox and a raised stone hearth across the

front. There would also be a warming bench and a woodbox on the kitchen side. The stove would be finished with grey river rock. I could see it. Standing in the dim light of evening, amidst bare walls and the smells of new construction, we shook hands and planned for a mid-October commencement.

Foundation and firebox—Working together, Miller and I started in the basement and laid up a foundation with 8-in. block and 12-in. by 12-in. pumice flue liner. We installed a cast-iron clean-out door and a 6-in. thimble, so that a stove

The stonework and glass door lend this hybrid masonry stove some of the charm of a traditional fireplace (facing page). The stove's thermal mass and system of flue baffles enable it to store and radiate heat over long periods of time. A heat exchanger installed inside the firebox also gives it the capacity for quick heating response. A circulating fan blows outside air through the heat exchanger and into the living room through openings over the fireplace door. Three cast-iron cleanout doors on the left side of the stove provide access to the flue baffles (left). Above the cleanouts is a handle for closing the damper. The back of the stove, facing into the kitchen, features a stone bench, a nook for wood storage, and a warming shelf (above).

could be added later, if the Millers decided to use their basement apartment.

Forming and pouring the slab for the firebox and hearth was no small task. We had to cut a main supporting beam for the floor, and cantilever part of the slab over foundation walls.

Also, we ran two air ducts between the floor joists from the garage to precise locations in the slab. One was a 1½-in. black iron pipe to carry combustion air to the firebox, and it needed to be on the centerline to connect to the combustor housing. Similarly, the 6-in. galvanized duct carrying cool air to the heat exchanger had to

fall just inside the front wall of firebrick and immediately behind the left wall of the firebox.

Miller is a structural engineer with the U. S. Forest Service in Missoula, Mont., and his specialty is bridge design. He designed the slab to carry 2,000 lb. per sq. ft. That sounded good to me, since the front wall of the stove would not stand directly above a foundation wall. The raised hearth would sit over the northwest foundation wall and the front wall of the stove would rest about 20 in. back, over an area of short span between block walls.

We began construction of the firebrick core

by measuring back from the flue 4 in. for the partition-block surround and another 1 in. for an expansion joint between chimney and stove. The back wall of the stove would rise along this line. We snapped a second line 27 in. into the living room from this line and parallel to it to establish the outside edge of the front wall.

The precast components I make are sized to be compatible with firebricks. Standard fireplace firebricks measure 2½ in. by 4½ in. by 9 in. My slabs, which create horizontal flue baffles, measure 2½ in. by 9 in. by 27 in. Their length determines the depth of the stove. With brick walls

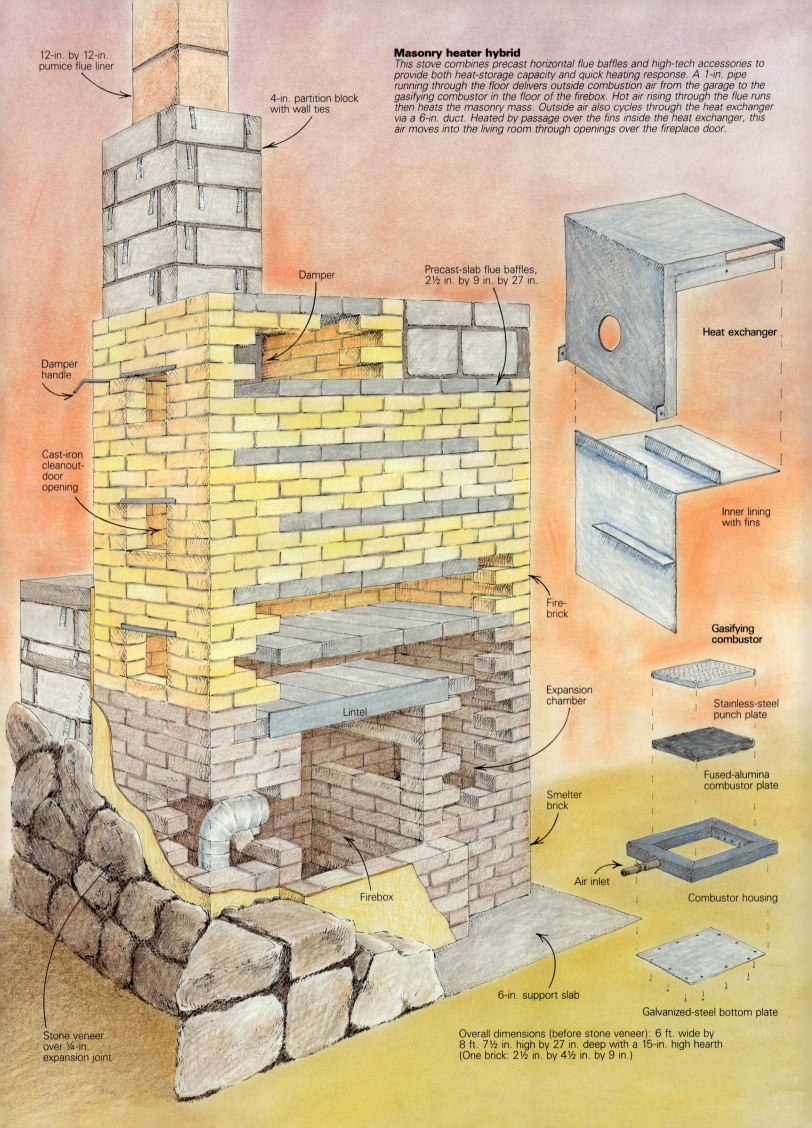

12-in. by 12-in. pumice flue liner

4-in. partition block with wall ties

Masonry heater hybrid

This stove combines precast horizontal flue baffles and high-tech accessories to provide both heat-storage capacity and quick heating response. A 1-in. pipe running through the floor delivers outside combustion air from the garage to the gasifying combustor in the floor of the firebox. Hot air rising through the flue runs then heats the masonry mass. Outside air also cycles through the heat exchanger via a 6-in. duct. Heated by passage over the fins inside the heat exchanger, this air moves into the living room through openings over the fireplace door.

Damper

Precast-slab flue baffles, 2½ in. by 9 in. by 27 in.

Heat exchanger

Damper handle

Inner lining with fins

Cast-iron cleanout-door opening

Fire-brick

Gasifying combustor

Stainless-steel punch plate

Expansion chamber

Lintel

Fused-alumina combustor plate

Smelter brick

Air inlet

Combustor housing

Firebox

Galvanized-steel bottom plate

6-in. support slab

Stone veneer over ¼-in. expansion joint

Overall dimensions (before stone veneer): 6 ft. wide by 8 ft. 7½ in. high by 27 in. deep with a 15-in. high hearth (One brick: 2½ in. by 4½ in. by 9 in.)

4½ in. thick, this makes the interior of the flue 18 in. from front to back.

We had agreed that for purposes of design and traffic flow through the house, the brickwork should be 72 in. wide. I like to design in multiples of 9 in. because this simplifies the construction. Since the width of the brick is exactly half its length, proper lapping, or "breaking," of joints can be accomplished easily with either full or half-length bricks.

Miller and I made the firebox larger than is typical for most masonry stoves so that more and bigger logs could be loaded at one time, and so that loading would be required less often. The firebox is 27 in. wide and will easily accommodate 24-in. logs. We extended the front wall of the firebox 4½ in. into the hearth area to enlarge the burning pit and to make the firebox easier to load and clean. This deviation gave us a firebox depth, front to back, of 22½ in.

The gasifying combustor on the floor of the firebox would sit under a pit of coals and drive off volatiles from the fresh load until the wood turned to charcoal. The stainless-steel heat exchanger and "big screen" glass door would provide immediate heat, while the masonry baffles absorbed and stored much of the rest.

The sidewalls of the firebox are a single brick thick. Since flue gases exit near the upper right corner of the combustion chamber, I shifted these walls 4½ in. left of center. This maximized an 18-in. sq. expansion chamber on the right side of the firebox while leaving a minimal 9-in. chamber on the left to house the heat-exchanger service duct.

With these features in mind, Miller and I snapped chalklines representing all the walls to make certain everything was where it should be. Then we thinned a 100-lb. can of Sairset refractory mortar (A.P. Green Refractories Co., Green Blvd., Mexico, Mo. 65265) to dipping consistency (creamy yet with enough body to form ¹⁄₁₆-in. mortar joints) and began setting bricks.

The dark bricks used in the lower portion of the stove came from the Anaconda Copper Mining Company's smelter in Anaconda, Mont. They are very dense, weighing 10 lb. each. By the end of the day, they seemed a lot heavier.

Miller and I stood on opposite sides of the fireplace with a tub of mortar between us, taking turns dipping first the bottom, then an end of each brick. Dipping the bricks, rather than troweling mortar, saves time and material. Mortar consistency is the key. Sairset ready for dipping is like thick pancake batter. We would gently mush the buttered end against the last brick, tap a few times until mud oozed out the joint and then grab another. A plumb line dropped at one corner of the stove gave us a starting point.

At the sixth course, along the left wall of the firebox, three bricks were hung out 2 in. to form a support for the heat exchanger. The L-shaped exchanger would cover the left wall and ceiling of the firebox. Its job is to heat cool, fresh air from the garage and blow it into the house through the vent over the glass door. An elbow was fixed to the 6-in. duct, and an opening made through the left wall of the firebox. High-temperature stove wires coiled in the duct connect ceramic switches in the heat exchanger to

fans mounted in the garage. One switch controls the combustion-air fan connected to the gasifying combustor. The other operates a 265-cfm cooling fan connected to the heat exchanger.

We stopped the front wall of the firebox at the sixth course, 15 in. above the concrete. The solid stone hearth would finish at this height.

In the rear wall of the firebox, I set a rectangular plug (about 9 in. by 18 in.) made of refractory cement. For times when little heat is needed (summer fires, for instance), the plug can be removed to allow heat and gases to bypass the upper flue baffles and pass directly up the chimney. For practical reasons, we should have fabricated a metal damper instead, operable from outside the firebox, which would have allowed the Millers to use it as a bypass for easy startup.

At the ninth course in the right sidewall of the firebox, we located the exhaust port through which hot gases and flames would pass into the expansion chamber and flue runs. The opening measures 7½ in. high, the height of three brick courses, by 18 in. long. Using refractory cement and crushed firebrick, I cast lintels and set them over the openings in both sidewalls of the firebox. When properly cast and cured, these components are rated safe by the cement manufacturer at temperatures up to 2,700°F. My stoves never see sustained temperatures this high, but there's something to be said for the peace of mind gained by overbuilding.

Flue baffles—We continued up with bricks to the fifteenth course, nine courses above the hearth level. I used an angle grinder with a carborundum blade to cut notches ⅜ in. deep in the top of the bricks at the upper front corners of the firebox. An angle-iron lintel 3½-in. by 3½-in. by ⅜-in. thick sits in the notches.

Next, a heavy bed of mortar was troweled on the bricks and the first row of slabs was placed. This course of slabs forms the roof of the firebox as well as the floor of the first baffle. Six slabs, side by side, create each of the five baffle levels in the stove, leaving a 9-in. by 18-in. opening at alternating ends of each run for the unrestricted flow of flue gases into successive baffles (drawing, facing page).

For cleaning access to the flue runs, I left 9-in. wide by 7½-in. tall openings in the brickwork, and covered them later with cast-iron cleanout doors. In this stove, three cleanout ports are located in the left end wall of the structure, centered at U-turns in the baffle system (photo left, p. 53). This allows easy access to both an upper and lower flue run through each port.

After the first row of slabs was in place we continued up with our walls. We ran out of smelter brick while building the first baffle and switched to common fireplace firebrick. We made all the flue runs in this stove 7½ in. tall (three courses). Each fourth course is a row of slabs that bind the long walls together and increases structural stability.

The fireplace cuts off a corner of the living room and divides it from the kitchen. Because it sits at a 45° angle to the room, the right rear corner of the stove actually stands in the kitchen. This meant that the fifth baffle, which finished off 4 in. below the kitchen ceiling, would

be the last full-length flue run we could build. A half-length sixth run at the left side of the stove would get us back to the chimney.

Where this corner of the stove extends under the sheetrocked kitchen ceiling, we laid a course of common brick at the stove's perimeter. Behind this brick, we sealed the tops of the baffle slabs with an inch or so of regular mortar, and covered this with 2 in. of loose vermiculite masonry insulation. We used partition block to fill in over the sixth baffle, leveling the right side of the stove with the short final flue run.

We installed a sliding damper, made locally of ¼-in. plate steel, in the last baffle. It rides in angle-iron tracks held firmly against the brick by stainless-steel wire. As a safety feature, we made the Millers' damper so that it doesn't close all the way; it leaves a ¼-in. space to allow carbon monoxide to escape should the damper be closed prematurely. The connections between stove and chimney were made with modified flue liners and refractory cement.

Although the physical labor was intense, building the firebrick core for this stove was surprisingly fast and simple. Including time spent packing materials, working mortar to the proper consistency and checking the drawings for construction errors, Miller and I built this structure in three working days. We used about 700 firebricks, 39 full slabs, 4 half-slabs and 200 lb. of refractory mortar. I figure the core of the Millers' stove, excluding the chimney and veneer, weighs close to 7,500 lb.

Through the roof—Once the core was complete, we turned our attention to the chimney and blockwork on the kitchen side of the stove. The Millers wanted a stone bench to the left of the chimney and a tall, narrow woodbox to the right (photo at right, p. 53). A second compartment over the woodbox was planned as a warming shelf for bread dough.

We placed a ⅝-in. sheet of rigid fiberglass between the chimney and stove as an expansion joint. This allows the stove brick to get hot and expand without affecting the structural integrity of the stack. Before starting the stonework, we covered all other vertical faces of the fireplace with ¼-in. thicknesses of this same rigid fiberglass insulation, except for the corners, where we went a little heavier with the material.

Portions of the chimney to be veneered were blocked up with hollow-core 4-in. partition blocks. Wall ties set in the block at 16-in. intervals connect the stone to the block. Partition block for the woodbox was worked into the right side of the chimney, laid against the back wall of the stove and returned into the kitchen 20 in. to remain flush with the chimney.

Two concrete slabs 3½ in. thick by 20 in. square were precast and set into the chimney/woodbox structure to form the warming shelf and its roof. One course of block filled with debris and sand became the base for the stone bench.

We were able to miss the valley in the roof with the chimney, but could not avoid cutting a doubled-up 2x12 beam at the north edge of the kitchen ceiling. The cut ends of this beam were temporarily supported by wood posts on hy-

After laying up the horizontal flue baffles and firebrick core, the author mocked up keystone arrangements over the fireplace to determine which stones would work best.

A stainless-steel heat exchanger covers the left wall and ceiling of the firebox. Outside air, driven by a circulating fan, is heated as it passes through the heat exchanger and moves into the living room through the vent over the firebox door.

Two ceramic sensors are mounted inside the heat exchanger. Wired with high-temperature stove wire, they serve as thermostats to control the circulating fans for the combustor and the heat exchanger.

Positioned to miss the valley, the stone-faced chimney rises 5 ft. above the roof.

An airtight glass door permits a view of the fire without compromising the heating efficiency of the stove. Openings in the bottom corners of the door frame allow secondary combustion air, preheated by the frame itself, to enter the firebox. Hot air from the heat exchanger blows into the room through the vent above the door.

draulic jacks. Eventually, both ends of the beam were supported by the stonework.

In the loft, a short section of stud wall had to be removed to let the chimney pass through. We used 4-in. solid block in this section because the back side of the chimney, located in an attic wing over the kitchen, was not to be veneered. Recent changes in the Uniform Building Code require a masonry flue to be wrapped either with an 8-in. hollow-core block or 4-in. solid block where no additional veneer is planned.

We ran block through the opening in the roof, flashed it and went up with 5-in. stone veneer placed directly against the flue liner. The chimney stands about 5 ft. above the roof and extends just over 2 ft. beyond the peak (photo facing page, bottom left). We finished the chimney on a beautiful, sunny Friday afternoon. By Monday, a Canadian cold front had blown in, signaling the end of Indian summer.

Stone veneer—Somewhere along the line, Brad, Cheri and I had taken a day and collected six tons of stone for the veneer, so when cold weather hit, we were ready. Of the many local stones I use, none looks more common on the ground or more subtly attractive laid up than the grey, pre-Cambrian quartzite the Millers chose.

We collected the stone along a dirt road in a narrow canyon where it had eroded off the parent rock and tumbled in the stream a short distance. This brief encounter with the stream rounded off the sharp edges but left the stone's angular nature intact. With this type of rock, it is relatively easy to obtain the look of washed river stone while maintaining tight, uniform joints.

Corners were no problem, either. Quartzites that aren't too massively bedded often break straight through the beds, providing an endless supply of 90° angles. Throughout the construction, we were able to use natural corners.

The colored mortar, deeply struck joints and fairly rough placement of stones are design features I've copied and embraced in my work as invaluable aids. A mortar joint darker than the stone it encases will often go unnoticed. A stark joint offends the eye.

We spent a lot of time striking joints. I buy new, flat, wooden-handled jointers and saw them off so the steel is 3 in. or 4 in. long. Twice a day we went over the rock we'd just put up, compacting more than cutting away, careful to leave a smooth, square face on the mortar.

Miller and I laid the six tons of stone that cover his fireplace in twelve working days. The creek stones weren't large enough to make decent seats for the raised hearth and stone bench. So we used massive rocks from the talus slopes that collect at the base of cliffs.

During the early stages of the stonework, we recognized a number of rocks spread across the floor as obvious candidates for keystones. We played with several arrangements on the floor, but what finally materialized over the fireplace door seemed almost to invent itself—a wonderful accident (photo facing page, top left).

On the other hand, the catenary arches over the woodbox and warming shelf were planned while picking the stone, and they seem somewhat forced, I think. Two long narrow stones

were set level with each other in the front wall to support a mantle to be seasoned and mounted at a later date. We capped the flat area over the warming oven and the top of the stove with 2-in. rock, overhanging the veneer a couple of inches, like the nosing on a countertop.

Heat exchanger and gasifying combustor—With the stonework complete, installing the high-tech accessories was the final step in what had been a long and exhausting construction process. The baffled, double-wall heat exchanger had been fabricated to our specifications in a local welding shop. And even though we'd been extremely careful with our measurements, we all breathed a sigh of relief when the unit was slid in and connected to its air-supply duct with no major hangups (photo facing page, top right).

The high-temperature stove wires coming from the fans were routed over the back side of the heat exchanger and run into the hot-air delivery opening. Ceramic sensors were wired in to automate the fans (photo facing page, middle right). The sensor servicing the combustion air fan turns it on at just over room temperature, and keeps it on until temperatures in the heat exchanger rise to 180°F.

The circulating-fan sensor kicks on its 265-cfm fan when the exchanger reaches 120°F and keeps it running until the air passing over the switch falls to 110°F. The inside face of the heat exchanger is 14-ga. stainless steel. The outside is a single piece of 14-ga. black steel. Three baffle fins $1\frac{7}{8}$ in. by 20 in. were welded to the stainless-steel liner (drawing, p. 54).

Primary combustion air comes from the garage through the 1-in. pipe to the gasifying combustor that is set in the floor of the firebox. The combustor is simply a dedicated air source that is located underneath the fuel load. It works together with the deep-pit firebox to generate high combustion temperatures and minimize thermal drag (which results in slow startup). The intense heat of the fire triggers a chemical change that converts molecules of the solid wood into a 2,200°F superheated gas. These gases move through the fuel load and "fire off" upon contact with secondary combustion air, introduced through the door frame.

Rectangular steel tubing (2 in. by 3 in.) mitered and welded into a 15-in. by 19-in. rectangle makes up the combustor housing. At one end of the housing, six ¼-in. dia. holes were drilled in the inner wall of the tubing. Next, the fused-alumina combustor plate (9 in. by 13 in. by 1 in. thick), which looks like a metallic sponge and diffuses the combustion air, was placed inside the combustor housing. A light-gauge piece of stainless-steel punch plate fitted over the combustor plate keeps fine ash from clogging the pores of the combustor. The bottom of the preheater is closed off with light-gauge sheet metal and self-drilling metal screws.

A low-cfm blower with the capacity to develop a relatively high static pressure is used in this system to deliver the optimum amount of preheated, low-turbulence combustion air for anticipated fuel loads. Seven to ten cubic feet per minute of primary air are supplied for an expected fuel consumption rate of approximately

7 lb. to 10 lb. per hour. Theoretically, a 70-lb. load should burn seven hours or more. In practice, we're seeing slightly elevated consumption rates due, we think, to air infiltration around the door and heat exchanger.

Tubular steel frames create the simple but effective fireplace door (photo facing page, bottom right). The jamb is made of 2-in. by 3-in. tubing, mitered at the corners and ground smooth before finishing. The door frame proper is made of 1-in. by 2-in. tubing and is attached to the jamb with a piano hinge. Japanese glass ceramic is sandwiched between rope gaskets and held against the inside of the door with stainless-steel brackets. A flat gasket fixed to the inside of the door frame compresses to form an airtight seal when the door is locked shut.

The turned wooden handle has a double catch for safety. One twist opens the door a crack and retains it, allowing enough excess air in to sustain intense blazes of short duration, the traditional method of firing masonry stoves. A second twist and the door swings open 90°, providing convenient access to the firebox.

The large frame against which the door closes doubles as a secondary combustion-air preheater, pulling room air off the hearth through openings at the bottom corners of the frame. Air ascends the vertical sides of the frame, fills the hollow tubing above and behind the door, and is forced down the glass through slots nibbled in the tubing. This feature encourages unburned hydrocarbons starved for oxygen to "fire off," producing beautiful, extended flames in a low-turbulence, high-heat environment.

The floor of the firebox was made with 2 in. of refractory cement over 3 in. of tamped sand. It finished out about 10 in. below the hearth, creating a firepit surrounded by four brick walls. This feature helps establish a deep coal bed while minimizing the ash removal.

Driving out the water—The fireplace was finished around the middle of January, 1986. Initially, it seemed to burn a lot of wood without storing much heat in the mass. However, after the first month or so, the amount of heat that was radiated back to the living space by the stone seemed more in keeping with the amount of fuel being consumed. This may have been due, in part, to the Millers' method of firing. Any high-mass, heat-storing fireplace takes some time getting used to.

But the real reason for poor heat efficiency during a stove's start-up phase probably lies with the water trapped in the masonry. Until the water in the structure is driven out by heat, which takes time with a stove this big, the mass can't possibly attain its true heat-storage capacity.

A four-hour fire in the uncharged mass will raise the surface temperature of the stone well above body temperature in twelve hours. Long after the fire burns out, the heat exchanger continues to cycle, producing hot air for two or three minutes at a time. The damper is never adjusted during a fire. Miller leaves it open, even between fires, closing it only when he leaves for extended periods. □

G. Karl Marcus is a mason in Missoula, Mont.

K'ang
An under-the-floor masonry stove

by Jeffrey J. Smith

Karl Marcus, a stonemason in Missoula, Mont., was experienced in building Russian fireplaces, and he wanted to include one of these masonry stoves in his passive-solar addition, a south-facing greenhouse/office. But Marcus didn't want to sacrifice any of the floor space (170 sq. ft.), which was at a premium in his older home. The solution was to build a masonry stove beneath the addition's floor.

Marcus's idea wasn't a new one but rather an adaptation of an ancient, multi-cultural tradition. In *The Book of Masonry Stoves* ($14.95 from Brick House Publishing Co., Inc., 3 Main St., Andover, Mass. 01810), David Lyle writes that the ancient Roman hypocaust and the Chinese

The k'ang firebox is small, 3.1 cu. ft., and designed to burn sticks rather than logs. But it must withstand temperatures of 1,100°F to 1,800°F. The firebox arches were cast in custom forms using refractory concrete.

k'ang, both subfloor heating systems, were used 2,000 years ago. An Afghan version, called the tawakhaneh, may have been around for as long as 4,000 years.

Marcus built his k'ang to wring every last Btu from his firewood. The stove does this by burning wood at high temperatures and by storing the tremendous heat that it generates. Marcus's k'ang won't need to burn a constant fire, either. He believes that an hour-long burn twice a day will be enough to heat the addition and contribute substantially to heating his entire house. There are 225,000 Btus in 30 lb. of firewood. If the k'ang is 90% efficient, it will generate and hold 202,500 Btus morning and night (see the sidebar on p. 61).

I had seen Marcus's work, and was intrigued by some other masonry stoves he'd built. I was especially impressed with the massive, under-the-floor stove in the conference center of the building where I worked. The heat it produced was very comfortable. So last spring, when Marcus mentioned that he had started to build his under-the-floor stove, I asked if I could watch the construction and write about the process.

Preparation—As soon as the ground thawed in the spring, Marcus began digging. At the center of the existing exterior wall, he excavated to a depth of 6 ft. below grade. He then laid in a concrete-block stairway, which began at the southeast corner of the house and dropped five steps to a 3-ft. square landing. All along the stairwell's interior wall, Marcus reinforced the

existing foundation with concrete. As shown in the drawing, the stairway turned left at the landing and dropped two more steps to meet the existing 3-ft. wide basement corridor. Marcus strengthened his foundation at this entranceway using bond-beam construction with each course of blocks all the way to the floor. He used blocks that had knockout webs so that removing the knockouts in each block created a horizontal channel through the entire course. Marcus then filled this channel with concrete. Next, he tore out the rickety old wooden stairway leading to the living room upstairs.

The concrete stairwell divided the new addition in half, and also framed the west wall of the k'ang stove. Three courses above the stair's landing, Marcus placed the k'ang's 3.1-cu. ft. firebox. He and his wife would be able to stoke the stove while sitting on the stairs.

With the stairway and the exterior frost walls in place, Marcus removed the house's south wall. He rebuilt 12 ft. of flooring to fill in where the back entrance and basement stairway had been. Now he was ready to build the k'ang.

Construction—The stove consists of poured concrete slabs, concrete blocks and refractory brickwork (drawing, below). On the west, it is framed by the stairwell; on the east and south, it is bordered by the concrete-block foundation of the new addition. Marcus first laid 2-in. rigid fiberglass insulation on the packed, leveled earth where he would pour his foundation slabs. He then poured a 3½-in. by 40-in. by 48-in. foundation slab for the stove's firebox. When it cured, he poured another 3½-in. thick slab to support the body of the stove. This 9-ft. by 8-ft. main slab overlapped the firebox slab on one end. Both slabs were mixtures of 3 parts ¾-in. crushed stone to 2 parts sand to 1 part portland cement.

When the Anaconda Copper Mining Co. in Butte, Mont., shut down its smelting operation five years ago, Marcus bought a pallet of new smelter bricks from them. These were top-of-the-line, extremely high-temperature bricks. Except for 50 store-bought refractory bricks that he laid in the south wall of the stove (the wall that would be stressed the least by the k'ang's heat), all of the stove's bricks were smelter

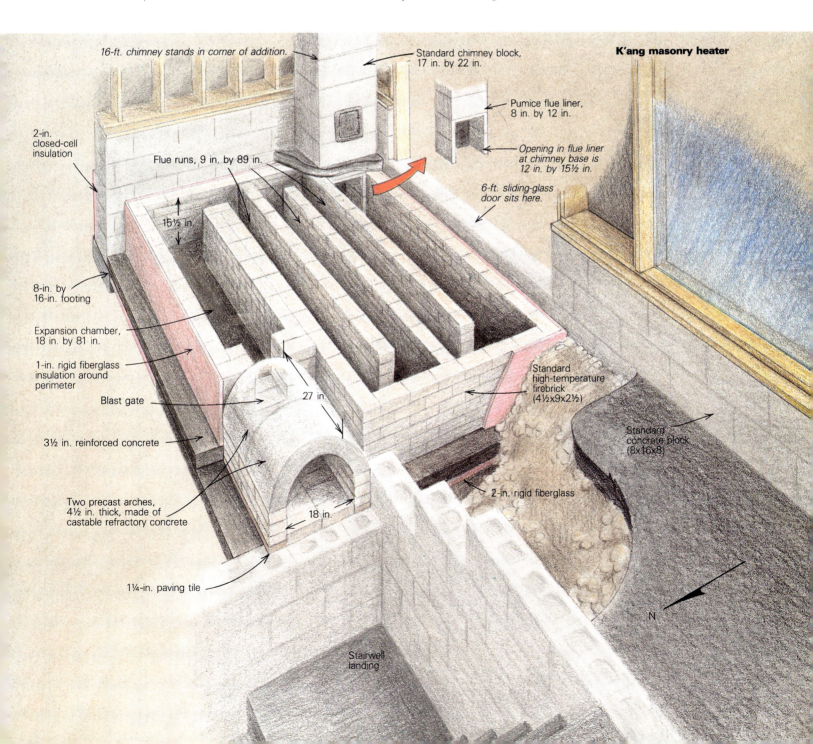

K'ang masonry heater

16-ft. chimney stands in corner of addition.

Standard chimney block, 17 in. by 22 in.

Pumice flue liner, 8 in. by 12 in.

Opening in flue liner at chimney base is 12 in. by 15½ in.

6-ft. sliding-glass door sits here.

2-in. closed-cell insulation

Flue runs, 9 in. by 89 in.

15½ in.

8-in. by 16-in. footing

Expansion chamber, 18 in. by 81 in.

1-in. rigid fiberglass insulation around perimeter

Blast gate

3½ in. reinforced concrete

27 in.

Two precast arches, 4½ in. thick, made of castable refractory concrete

18 in.

1¼-in. paving tile

Standard high-temperature firebrick (4½x9x2½)

Standard concrete block (8x16x8)

2-in. rigid fiberglass

N

Stairwell landing

Like the firebox arches, the lids of the first three flue runs were also precast with refractory concrete. To allow for extra expansion and contraction due to the high temperatures, Marcus used nine separate sections to make the lid of the expansion chamber.

bricks. Marcus used Sairset refractory mortar manufactured by the A. P. Green Refractories Co. (Green Blvd., Mexico, Mo. 65265).

Flue runs and firebox—Next, Marcus laid out the elaborate interior flue structure, a labyrinth of vents and channels designed to absorb the firestorm created in the firebox. The fire pours through a 9-in. by 9-in. blast gate in the firebox's double-brick rear wall and enters the first flue run. The blast gate acts as a baffle, creating turbulence and burning up exhaust gases.

The first flue run must withstand great temperature fluctuations as well as occasional sustained, fiery blasts. It's called the expansion chamber because the gases and smoke are actively combusting, and therefore expanding, as they leave the firebox. Marcus made the chamber 18 in. wide—the same width as the firebox and is double the width of the succeeding flue runs. It is 15½ in. tall and runs 81 in. along the north wall, which is insulated with 1-in. rigid fiberglass. This is the only wall backed by packed earth. The interior wall is a double brick wall.

At the end of the expansion chamber, the hot blast collides with a brick wall one brick thick backed by 1-in. rigid fiberglass and the concrete block foundation of the room's east footing. The exhaust escapes through a 9-in. switchback into the next flue, which is also 9 in. wide. This second chamber is 89 in. long and ends in another switchback to the third flue run. The second, third and fourth chambers are identical in width, height and length. The final chamber is 13 in. wide and ends in a standard 17-in. by 22-in. pumice-lined chimney.

Marcus custom built the firebox's twin 4½-in. thick arches (photo, p. 58). The inner (bottom) forms are plywood and stretched galvanized steel. He put a piece of ³/₁₆-in. plastic on top of the curved form to keep the concrete from ad-

hering to the steel. The form's outer (top) piece was made by a local plastics manufacturing firm and was designed by Marcus to be held in place with clamps. He poured Refracrete (North American Refractories Co., 1500 Houser Way S., P.O. Box 975, Renton, Wash. 98055), a large-aggregate, castable refractory concrete, through an opening at the top of the arch, and the pieces cured in the same positions they would take above the firebox. The forms are reusable.

Marcus then built forms and poured a series of 2½-in. thick lids for each flue run. For the lids of the first three chambers, he used Refracrete. Marcus poured nine separate sections (9 in. by 27 in.) to cover the expansion chamber (photo above), thinking that the joints would allow for the expansion and contraction due to the high temperatures in this first flue run.

For the second and third chambers, he poured 13½-in. wide slabs that would cover all but 27 in. at the east wall. He needed cleanout ducts there, and he poured these duct lids separately to include 8-in. dia. holes that would accept airtight, sheet-metal lids (top left photo, facing page). The lids that covered the fourth and fifth flue runs Marcus cast from the 3-2-1 portland mixture he'd used for the slabs. He added a small amount of Mason's Blend, a fireclay made by North American Refractories, to the lids to increase their elasticity and heat resistance.

Final steps—When Marcus had mortared the seams in the flue-run lids and had filled in the sides of the arches with vermiculite, he laid a gridwork of rebar over the surface of the stove. He installed the two airtight cleanouts, one at the start of the second flue run and the other between the third and fourth chambers.

Marcus also installed a 3-in. dia. aluminum pipe across the top of the flue runs and insulated it with fiberglass (top left photo, facing page). He

was hoping that a 165-cfm fan would blow cold basement air through the pipe to two small ducts along the south wall, where the heated air would emerge and circulate through the house. This turned out to be ineffective. Marcus has since installed a small fan at floor level on the west wall; it simply blows air across the floor, up the wall and back into the house.

There is a carbon-monoxide danger associated with masonry stoves. Some European countries, most notably Switzerland and Austria, have a regulation that chimney dampers on masonry stoves cannot close more than 35%. Marcus is convinced that his k'ang will draft rapidly enough and is effectively sealed against the escape of exhaust gases. But as an extra precaution, he has no damper in his chimney.

When all this was done, Marcus poured a 4-in. concrete slab over the stove. It was made up of the 3-2-1 portland mixture with a small proportion of fireclay. He dyed it black for greater absorption of winter-time solar heat (top right photo, facing page).

The 16-ft. chimney was the last step, and it went up without a hitch. Marcus added an airtight, 8-in. by 8-in. cleanout door just above the floor. His business, Native Rock Design, got busy before he could build and install an airtight firebox door. But that didn't prevent him from christening the k'ang with its first roaring fire. There was no hesitation in the draft, and once established, a pocket of white-orange flames jumped through the blast gate to feed hungrily on exhaust gases to the end of the expansion chamber. Marcus brought out the Irish whiskey to celebrate, and I wished I'd saved one final frame of film to capture his satisfied grin. □

Jeffrey J. Smith is a Montana-based writer with a strong interest in natural-resource issues and energy efficiency.

After the flue-run lids were complete, Marcus laid in rebar and installed two 8-in. cleanout ducts with airtight lids (left). He also laid in a 3-in. aluminum pipe wrapped in fiberglass that was intended to heat cold air from the basement. Marcus added black dye to the concrete for the finished slab (above) in order to increase the absorption of winter-time solar heat. The k'ang, finished except for an airtight firebox door, is shown below. Photos above: Jeffrey Smith.

Metal stoves vs. masonry stoves

All masonry stoves are designed to have enough mass to store heat and deliver it slowly, over a long period of time. This contrasts sharply with the idea behind cast-iron woodstoves, which are now being used by more than 20 million Americans.

Most cast-iron, airtight woodstoves burn large loads of wood for quick response. Their thin airtight walls radiate the heat into your room. The warmth is stored in your walls, hearth, furniture, carpet, even in you if you linger near the stove. It is your room, then, that heats the air. Since most rooms lack masonry-mass heat-storage systems, you must maintain a fire. To keep the fire from burning too hot, you have to crank down the stove's damper, which deprives your fire of air and causes it to smolder.

Though it does the job for up to 12 hours without refueling, a cast-iron stove wastes much of the wood's energy. Researchers at Auburn University and at the New Mexico Energy Institute have found that one-half to two-thirds of the fuel value of seasoned firewood is in gases and volatile liquids.

The key to using those gases and volatile liquids is high combustion temperatures, 1,100°F to 1,800°F. But many cast-iron woodstoves actually begin to glow red at less than 900°F. Sadly, up to 50% of

the fuel you've paid for will vanish up your chimney. And that's the same unburned fuel that is loaded with creosote and air pollutants.

Some cities, like Missoula, Mont., lose a half-dozen residences each winter because of chimney fires, and their officials are beginning to restrict wood burning. When the Missoula health department, for instance, finds more than 150 micrograms of respirable wood-smoke particulate per cubic meter of air, they require everyone to switch to fossil-fuel sources of heat. For 10 days last winter, Missoula's wood burners had to shut down their stoves because of air pollution.

Some countries have long traditions of clean-burning masonry heaters. Two-thirds of the new homes in Finland are built with masonry woodstoves. The Finnish government even gives tax breaks to encourage their use.

Also, though few tests of masonry-stove efficiency have been performed in the United States, European tests have placed masonry-stove efficiency at 70% to 90%. That means that almost all of the heat value of the wood (Btus) is used in a well-constructed masonry stove. The fires in these stoves reach 1,100°F in the first three minutes. After an hour, they reach 1,800°F.
—J. J. S.

From Boulders to Building Blocks

How a traditional stonemason quarries and dresses sandstone

by Charles Miller

Benny Soto doesn't have to look at the chisel anymore when he dresses a block of sandstone. His hammerhead instinctively finds the butt of the chisel, sending a steady clink, clink, clink ringing around the building site. The stone chips fly about, as he transforms another ordinary rock into a hand-tooled flagstone.

It wasn't always this easy for Soto to hit the chisel butt dead center. Sixty years ago, when he moved to Santa Barbara, Calif., from his native Guadalajara, Soto started his masonry career by lugging stones and digging ditches for a group of Italian stonemasons. Sensing that he had more to offer than a strong back, his boss urged him to learn the stonemason's trade. Soto agreed to give it a try—anything had to be better than lifting and toting rocks about all day, broken only by bouts of ditch-digging.

But the shift from hard labor to skilled craft wasn't without difficulty. Many of Soto's unpracticed mallet strokes hit the chisel butt slightly off center. The hammerhead would glance to the side, and the big knuckle on his left hand would take most of the shot. He hit his hand so many times that he developed blood poisoning, and he nearly lost his resolve to learn the trade during the two weeks that it took him to recover. But the thought of going back to the ditches was a powerful incentive, and Soto stuck with it.

The on-site quarry—Over the last 60 years, Soto has built walls of random rock, flagstone patios, fireplaces with squared-off sandstone blocks and baronial entryways topped with S-curved capstones. He quarries the stones himself, and given Santa Barbara's notoriously rocky soil, he usually needs to go no farther for raw materials than any nearby foundation trench. Some of these virgin stones are the size of beach balls, others are as big as hippos. The big boulders are easiest to work, for the same reason that you get more uniform slices from a loaf of bread than you would from a biscuit.

Although electric and pneumatic drills and chisels are now available, Soto relies on the kinds of tools that stonemasons have used for

Reducing a boulder into building blocks begins with cutting it in half, then dividing the sections into ever smaller pieces. In the photo at right, Soto uses a lifter to start holes for the wedges that are used to split the stone. Once the wedges are in place, he drives them into the stone with a sledgehammer. The wedges have to be hit alternately to ensure a smooth cut.

centuries. He thinks that handmade work should look handmade, and power tools (besides being too noisy) take away some of the artisan's control. Soto's tool bucket contains steel wedges, cold chisels and a 4-lb. hammer (photo right). The hammer has a hickory handle, which absorbs some of the shock of hitting the chisels. If the handle gets slippery, Soto roughens up the hickory on the edge of a stone. If a handle breaks, he shapes a new one to the right contours, using a piece of broken bottle as a scraper.

The chisels are of four varieties: pitching tools, points, lifters and toothed chisels. A pitching tool looks a little like a brick chisel, but its cutting edge is blunt. It's used to whack off pieces of stone near the edge of a block. A point is a cold chisel with a tip that's about as sharp as a railroad spike. It's used to excavate the holes needed to split the stones, and to dress the stone. A lifter resembles a point with a blunt tip, and Soto uses it primarily to begin the slots in the stone that will accept the wedges. The toothed chisel creates a texture on the stone's surface that resembles cross-hatching.

Most of the tools that Soto uses are available commercially (The Bicknel Co., P.O. Box 627, Rockland, Maine 04841, and Trow & Holden, P.O. Box 475, Barre, Vt. 05641 are two sources). He has his wedges made by a blacksmith.

Sandstone—Like limestone and shale, sandstone is a sedimentary rock. Sandstones are held together by various kinds of naturally occurring cements. The yellow and reddish versions indicate iron-oxide cement. Other types can be white, black, cream-colored or even green. When sandstone breaks, the fissure usually opens through the cement, rather than through the grains of sand. This property makes sandstone relatively easy to shape.

When Soto sizes up a rock that he is about to break into building blocks, he thinks about waste. How can he best use the rock with as little waste as possible? Soto is adamant on this point, and tries to put every offcut to use. He won't, however, reuse stones that have previously been in contact with mortar. Elements in the mortar evidently leach into the sandstone, making the stone brittle and unpredictable to cut. Soto says such stones are dead.

Quarrying a sandstone boulder is a matter of reduction. A large rock is cut into ever smaller pieces and eventually into usable blocks. Some sandstones have a grain to them, and the first cut should follow it. Typically, a boulder will be quartered (drawing, right) and the dimension of the slices taken off the quarter-sections will be determined by the task at hand. If, for instance, Soto is making fireplace veneer blocks, which are about 18 in. long, 9 in. high, and 5 in. deep, he will make sure there is a usable 20-in. thick portion in the next slice he takes off the boulder. The excess "meat" is an allowance for a slightly erratic cut—it can be easily trimmed away when the stones are dressed. If the cut goes radically awry, chances are he will still end up with a piece of stone that has usable dimensions. If he tried to carve off a 5-in. thick piece and failed, it's likely that little of the material would be salvageable. Also, it's easier to get

Clockwise from the top, Soto's hammer, a point, two pitching tools, a toothed chisel and a lifter. The three wedges were custom made by a blacksmith.

straight cuts when there are roughly equal amounts of stone on both sides of the cutline. Much of this quarrying process is guided by an intuition that comes only with experience.

Once he has the 20-in. thick piece lopped off, he cuts it in half again. If the stone co-operates, he may now trim off the waste portion near the curved edge. If he's in doubt about the accuracy of this cut, he will split out the blocks and trim them individually. In this manner, large boulders are cleaved until they are reduced to blocks that are about 20 in. by 10 in. by 6 in.

Making the cuts—Soto begins a cut by using a point or a fat, soft pencil to mark a line on the stone. If the stone is still in the round, the line he makes is across the top of the stone, and it is straight in plan. Soto's straightedge is an ancient length of 2x2, and if he needs to square it with another line he relies on his eye.

He uses a lifter to begin a series of wedge slots along the cut line, as shown in the photo on the previous page. The slots are 3 in. to 4 in. on center, and never closer than 2 in. to the edge of a stone. When each slot is about ½ in. deep,

Quarrying a boulder
Large boulders yield the best blocks with the least waste. They are typically cut in half, and a section is levered onto its side and again halved. Then slabs of the appropriate dimension are cut away from the quarter sections and reduced, roughly by halves, to the desired blocks.

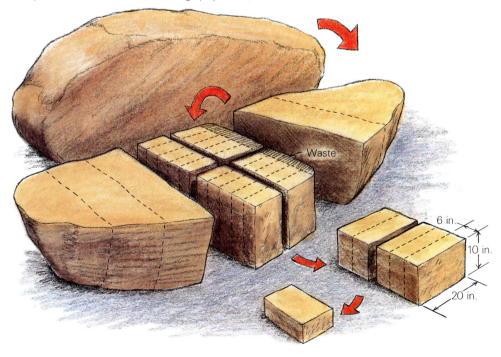

Waste

6 in.
10 in.
20 in.

The weighty presence of hand-wrought stonework is entirely in keeping with the sturdy detail in this Spanish Colonial Revival style home. Soto used a plywood template to regulate the curvature on the bottoms of the corbels that support the mantel, and he shaped them with a pitching chisel.

If opposite faces are not in the same plane, mark sides with parallel lines and remove excess stone with a pitching tool.

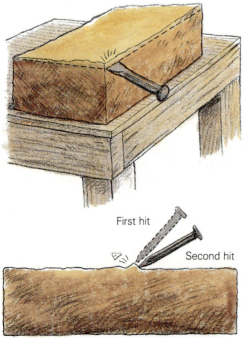

First hit

Second hit

To remove surface projections, use a point held at about 45°. Lower the angle for stubborn bumps.

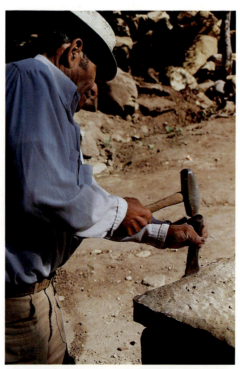

A pitching chisel is used to remove unwanted material in a hurry. Here Soto uses one to clean up a ragged edge on a flagstone.

Soto switches to the point and excavates the slot another ½ in.

Now the wedges are inserted into the slots, and Soto methodically drives them, alternating from one wedge to the next, with blows from a 16-lb. sledge. Soon a fissure opens, and the stone fractures in half. If the stone is a big one, he uses a long prybar to lever one of the halves on its back.

Stone dressing—Once he has a pile of rough-cut blocks on hand, Soto takes them one by one to his work table, a sturdy platform made of 2x6 braces, 4x4 legs and a ¾-in. plywood top. It measures about 3 ft. square, and its height is about 6 in. below Soto's beltline. With a stone on the table, he can hold his tools at a comfortable, waist-level height without having to bend.

If he's making blocks that need regular dimensions, Soto will check the block for twists or out-of-square corners. A straight 2x2 is used for the twist test, a framing square for the corners. If a block needs trimming to bring opposite faces into the same plane or to straighten an edge, Soto marks the stone accordingly. Removing this unwanted material is the pitching tool's job. With the curved back of the tool on the side opposite the workpiece, Soto cleaves away unwanted stone with sharp raps from the hammer (photo below left). It is the pitching tool that gives the edges of the blocks the broad facets that make handhewn stone so attractive.

Any bumps and projections on the face of a block are removed with the point (drawing, below left). Soto makes this work look effortless, with the tip of the point finding the base of a projection a millisecond before the hammer strikes the butt of the chisel. Stubborn bumps get two or more hits, the first with the point held at about a 45° angle, and subsequent shots with the angle approaching 30°.

If he's making flagstones, Soto doesn't have to worry about square corners and parallel edges. Instead, the task is to make the stones as flat on one side as possible, and then finish them with a pleasing texture that won't get slippery after years of use. For this he uses a tool called a bush hammer, which looks something like a meat tenderizer. Soto's bush hammer weighs 5 lb. and has 36 teeth on its face in six parallel rows. Soto lets a helper use the bush hammer, which has to be pounded over the entire surface of the stone within about 2 in. of the edges—any closer and the stone is liable to break. This work is for bruisers—15 minutes on the bush hammer will make your forearms blow up like Zeppelins. After hammering, the stone is swept clean with a stiff bristle brush to reveal a pleasing stippled texture.

Many of Soto's clients hire him to craft fireplace surrounds, hearths and mantles (photo above left). Whenever he does a fireplace, he cautions the mason who installs the pieces to use a stiff mortar mix, and thereby avoid messy drips that could discolor the stone. If some mortar does get on a stone, he recommends cleaning it with a stiff bristle brush. Dip the brush in water, shake off the excess and run it over the mortar stain, but in only one direction. Back and forth will drive the stain deeper. □

Maine Stonework

A personal approach to laying up and pointing rock

by Jeff Gammelin

My stonemasonry is inspired by the ledges and outcroppings of coastal Maine, and my approach is the result of studying these natural formations, in which the stones lie upon each other with at least one point of contact. The joints are tight, and I try to create, through a minimal use of mortar at a given point, the feeling that there is at least one spot where each stone rests directly upon another. This point of contact is one of the basic distinctions of my stonework.

When my wife and I began building with stone in our home, we realized our focus on making a collection of individually beautiful stones was shortsighted. The stones by themselves were beautiful, but more important, they needed to complement each other with regard to shape, color, texture and size. Using a wide variety of stones, from string-bean size to 500-lb. mantel stones, provides a contrast that suggests the interdependence so evident in natural rock formations.

My crew and I use freeform arches and serpentine lines within the mass because they imply strength, and we accomplish a feeling of

Stone is an elemental part of nature. It is the stonemason's task to take stone from its natural context and to use it in forms that participate harmoniously with their surroundings. This translation from the natural to the controlled reveals the individuality of the stonemason's art. Every mason I've seen lays stone differently. My methods spring from the style of my work. For example, I want to keep the individual stones as natural-looking as possible once they are laid. To do this we begin by laying up dry a small group of stones using small rock chips as shims. My crew and I work mostly with granite, and a bit with basalt and sedimentary stones. If stones need to be shaped, we do it with carbide-tipped points (9-in. long steel bars with a point on one end and a flat for striking on the other), hand

sets or chisels. Large granite stones that need to be worked for steps, mantels or lintels are cut. This is done by snapping a line, tracing it several times, drilling 2-in. deep holes every 4 in. to 8 in., and inserting half-rounds and wedges in the holes. The wedges are driven until the stone breaks. A hand set and point are used for trimming. We soften the resulting sharp edges by rubbing or crushing them with the stone hammer or by filing them with a rasp. Hammer marks are hand-rubbed to darken them.

Before we lay a group of stones, we often mark each one with chalk to show its position relative to the others, and to show how far the mortar line should extend to the face of the stone. This helps us control shadow and texture, and is useful on corners and in niches.

Mortar and pointing—We usually use dark masonry cement which, along with the varying degree of horizontal stone surface exposed, produces an interplay of shadow and texture. We work with about five different viscosities of mortar, depending on temperature and humidity, the type and size of the stones, and on whether it's interior or exterior work. Basically we like a rich, plastic mortar—wet enough for a thin vertical application but dry enough to support the stone. A good understanding between mason and tender is important to a good mix. Making a batch of mortar ideally suited to the conditions takes a lot of practice, but the influence it has on the smoothness of the operation cannot be overestimated.

The mortar is spread just thick enough to sup-

restful substantiality by using strong, solid-looking stones at the corners.

Our stonework falls roughly into two categories: composition within a rectangular framework, and the creation of more organic or sculpted forms. Working within the rectangular form of a fireplace is like painting on blank canvas. The composition proceeds within definite limits.

We try to develop a mosaic quality, blending size, shape, color and texture into a composition in which the stones have a life and character that derives from their participation in the whole. We concentrate on two things in establishing the design. First, because the observer's eye starts by moving along the joint lines, we manipulate joint depth, and the continuity, size and articulation of the joint lines. Second, because the eye soon moves along the stone masses themselves, we place individual stones to create larger forms. These forms should have an organic quality that reflects the nature of the material. The details of the stonework should be mutually dependent and intrinsically related.   —J. G.

port the stone evenly at all intended points. A tighter joint is generally a stronger joint. The less the stone is moved around after it is laid in mortar, the neater the joint and the less likelihood the mortar will run water over the face of the stone. It's important to prevent mortar from running onto any surface that will be exposed in the finished wall or chimney. While cleaning with muriatic acid can rectify some mistakes, the overall impression is sharper and more controlled if messes are avoided. After the group of stones is nestled into place, it is left undisturbed for a few hours until the mortar in the joints becomes somewhat leatherlike.

We then tool the joints with a thin piece of flexible steel. We usually don't add mortar, but we sometimes scrape it away in small areas,

making sure it is well compressed to seal the stonework. Exterior stonework requires more attention, because we want to be sure we eliminate all of the recesses or cavities that could trap moisture.

Our procedure is in contrast to the common practice of pointing at a later date by adding a differently formulated pointing mix. I'm critical of that method for two reasons. First, the pointing mix is usually brought out to the face of the stone, and cups over a bit of the stone's exterior surface. There's a great danger that this mortar, exposed to the elements, will lose its seal with the stone and direct water behind the outside surface. The freeze-thaw cycle can result in spalls or cracks that may need attention.

Another, more serious problem is that if the

wall is going to be pointed later, the mason usually doesn't pay much attention initially to forming a good, thorough seal between the stones. In fact, the mortar is usually left coarse so that the pointing mix will adhere. So once water penetrates the outside surface of the structure and is directed to the space between the stones, there is little to prevent it from traveling deeper into the work.

With our method, the pointing is neater, quicker and ultimately stronger, because it is integral with the mass. And because we point up each day's work before we pick up and go home, problem areas are dealt with while they are still fresh on our minds.   □

Jeff Gammelin lives in Ellsworth, Maine.

Repairing Old Stonework

With the materials already delivered and paid for, why not use them?

by Stephen Kennedy

At first glance, a lot of old stone buildings and foundations look like ruins ready for bulldozing. But much of the work that went into those structures was in gathering the stones and bringing them to the site. In the case of the pre-Civil War barn foundation pictured below, we're talking about 135 tons of carefully selected fieldstone, some of which is rough dressed with hammer and chisel. That's an incredible head start, even if all the stones need to be relaid. And luckily, three-quarters of the stones in this foundation were still in the right position, needing only to be repointed.

The major problems were created by the rotting of the wooden lintels over the windows, which caused the stone above to collapse. Making matters worse, the top-plate timbers had decayed so thoroughly that the resulting humus supported vegetation whose roots had damaged the top course of stones.

The relaying and repointing that we did on this foundation is fairly typical of stonework repair, and seems a good occasion to note some guidelines for such work.

Bed first, point later—There are several reasons to bed stones and point the wall as separate procedures. For one, you never have to work above the finished project—you build first, then point your way down the wall without leaving a mess. By laying up and pointing down, you

Color photo: Will Lane

The wooden lintels over the windows of this pre-Civil War barn foundation had rotted (below), causing the stonework to collapse. Though it looked like a ruin, most of the stones needed only repointing. With new lintels and all the joints neatly packed with fresh mortar (above), the foundation once again supports a structure.

also avoid shocking the finished joint with big stones being set above. It is much less crucial when the bedding mud gets jarred and cracked from new work.

If you take this approach, you can use different mixes for bedding mortar and pointing mortar. This will allow you to fine-tune each mix for the different jobs each has to do. Another advantage of a separate bedding mix is that variables in the weather are less likely to halt production. If you go strictly by the masonry books, only about three days a year here in Pennsylvania are fit for laying stones. It's either too hot or too cold, or there's a chance of frost or rain. With the separate bedding approach, you can work through the bad weather, and try to do the pointing during the few good days.

Weaker is better—The bedding mix in a stone masonry wall needn't be very sticky or have much compressive strength. A typical cubic foot of stonework weighs around 144 lb. and exerts a force of one pound per square inch (psi) on the mortar below it. At the bottom of an 8-ft. wall, the force increases to 8 psi. A two-story building on top of the wall might only double the weight. There is simply no advantage in using a mortar mix that will support thousands of pounds per square inch when all you need is 100 psi at the most. There's no sense wasting portland cement where it isn't needed. In fact,

there are disadvantages in using bedding mortar that is too strong.

Excessively strong bedding mortars (high in portland cement) contract on setting, causing cracks that open passageways for water. Also, a mortar that is nearly as hard as the stone will not cushion the stones from movement due to settling or to expansion and contraction. The walls that last longest do so because they can handle a certain amount of motion. When it is stressed, the mortar will give, sparing the stone.

Weak mortars (low in portland cement) allow the passage of water vapor and accommodate changes in humidity in a wall. Waterproof masonry coatings, which prevent the passage of water vapor, can cause spalling on the surface of walls. Water needs to be kept out, but a good pointing job will do that. Water vapor should flow more freely. More significantly, mortar high in lime can react with water and carbon dioxide in the atmosphere and seal its own fissures. The longest-lasting mortar joints have no portland cement in them.

Masonry manuals abound in contradictions, but most agree that sand grains of varying size will best stretch your cement and water, giving you a stronger mortar for your money. They also stress the importance of using clean, drinkable water in the mix in order to minimize organic particles, though I feel confident that I could mix a bedding mortar, using dirty water, that would last for centuries.

The temperature of the ingredients significantly affects the way any mortar sets, and bedding mortar is no different. It's best to keep all the components between 40°F and 80°F (50°F to 70°F is even better). Colder temperatures generally cause slower setting, and warmer ones speed things up. In summer, you can get into trouble by using water that's been heated in a long hose lying in the sun. Mortar that freezes before setting is also pretty worthless.

The bedding mortar we used on this job was (by volume) 1 part Type 1 portland cement, 2 parts Type S hydrated mason's lime and 13 parts regular mason's sand. We added water gradually to the dry mix until we had a workable paste. The amount of water needed depends on how wet the sand is. Soaked sand often requires no additional water for mortar.

Setting the lintels—The owner of the barn wanted to repair its foundation and build a new house on top. So he and I tore down the old stonework to the level of the window tops, except for the corners and an area between windows, which we left stepped up to build into later. We had the dubious good fortune of acquiring a pair of 1,200-lb. granite blocks for replacement lintels.

When we were ready to set the lintel stones, we enticed some bystanders into being slaves for an hour. They demanded to see the cold beer first, but then helped us build crude ramps from the ground up to the wall (photo above right). It's important not to rest ramps right on the wall because the top stones can tip dangerously. Also, planks can get trapped under the lintel if they extend onto the wall.

With five of us pushing from behind, we slow-

On the promise of cold beer in return for their brawn, a handful of bystanders agreed to help flop the granite lintels up the ramp to the windows.

Keeping it together

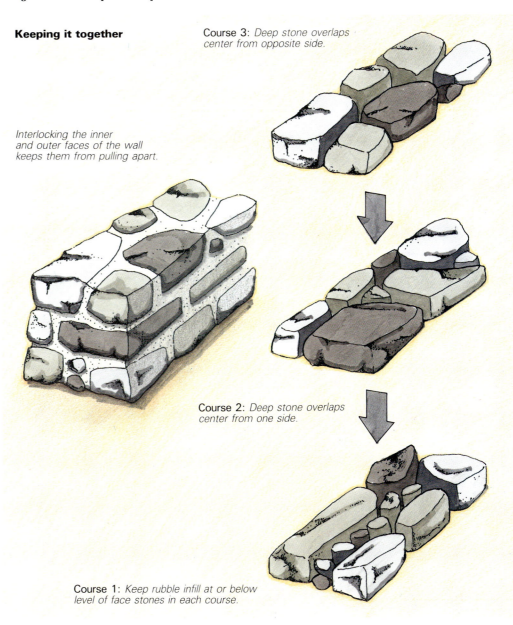

Course 3: *Deep stone overlaps center from opposite side.*

Interlocking the inner and outer faces of the wall keeps them from pulling apart.

Course 2: *Deep stone overlaps center from one side.*

Course 1: *Keep rubble infill at or below level of face stones in each course.*

ly rolled a lintel up the ramp. When it was one flip from the top, we adjusted its position so that the best face would end up on the outside of the wall. It's much harder to move the tonnage once it's sitting on the wall.

When we flipped the mammoth lintels into place, the stones beneath were displaced and their bedding mix was broken up. Using a car jack, we raised the lintels a few inches, repaired the damage and pushed in a fresh bed of mortar. The new bed was slightly stronger than the usual mix because of the increased pressure on those areas (I would likewise use a stronger mortar on either side of an arch).

Putting the old stones to bed—With the new lintels in place, we started building the wall back to its original height. We laid up the ends first, then filled in the middle. If the stonework in the middle goes up first, it dictates how you make the ends, which are far more critical.

The same is true for the thickness of the wall. The 2-ft. thick walls of this foundation are essentially two separate walls, with rubble infill. If the middle gets above the faces, it sets limits on the choice of the next face stone.

Interlocking the inner and outer faces of the walls by overlapping the stones across the center keeps the wall from pulling apart. This overlapping is best done by continually creating flat surfaces on which to work. Then it's a matter of laying the deeper stones on alternate sides in successive courses (drawing, previous page).

Turning raw, uncut stone into solid walls is a very slow, conservative process. Here are a few guidelines that will help make the job easier. First, laying up the stones in courses is essential to getting well-crossed joints. The height of each course is determined by the end stone at a corner. The middle stones needn't all be the same height—they can be added together in various combinations to bring the course up in layers to the desired height.

For appearance and structural integrity, spread out the big, beautiful stones. Beginners often reach for all the best stones right away. But you should avoid the syndrome of the beautiful lower right-hand corner. Most authorities on the subject say to lay the big stones on the bottom of the wall, using smaller ones as the height of the wall increases. I disagree. The top of the wall is most vulnerable to being knocked loose, and having big heavy stones there will do a lot to protect it, especially if the wall has no building on top.

Individual stones should be laid in their most stable position (i.e., flat). I can't build a good wall if the stones aren't approximately rectangular, although thicker walls can more easily accommodate irregularly shaped stones. In the balancing act of rubble masonry, the idea is not to try anything fancy.

As we built these walls, we roughly scraped back the bedding mix before it set up to leave room for an inch or so of pointing mortar. Smears of bedding mix can be removed with a wire brush and water, as shown in the photo below. It helps to have a hose handy. If cleaned within 24 hours, this low-in-portland mix comes off very easily. I usually spend about ten minutes first thing each workday cleaning the mess from the previous day's laying. This good soaking also ensures that the mortar will cure more slowly. Additional soakings should be applied regularly for about a week.

To the pointing—On this foundation, after all the stones were laid, the bedding mortar raked back and the wall cleaned, we were ready to point the work. Like a continuous bead of permanent caulk, pointing mortar keeps the bedding mix from eroding and spalling. Pointing mortar must be harder and stickier than the bedding mortar in order to endure the weather.

Other things being equal, mortar acts differently as its thickness changes. Joints narrower than ½ in. don't have much integrity unless finer sand and a richer mix are used. Fat joints do better with mortar that has more and larger sand. Round sand is good, perhaps preferable, if it is clean and graded, and if extra lime is used in the mortar to compensate for the decreased adhesion or friction between the grains.

When repointing a stone wall, there is a limit to how large a lump of mortar you can keep in place. A fist-sized blob will often slump out. I usually fill large spaces with small, clean stones. Deep holes may need to be filled in layers.

The pointing mix we used was (by volume) 1 part Type 1 white portland cement, 2 parts Type S hydrated lime and 9 parts mason's sand. Using white portland cement not only makes a lighter-looking joint, which looks more like the high-lime joints in older masonry, but also makes smears easier to remove. If you use pigmented mortars, it is advisable to start with white cement and sand, as the grey of regular portland is difficult to cover. White portland gives a slightly weaker mortar, but with laid stone walls, that's not a problem. Again, compressive strength is not as important as the flexibility that high-lime mortars achieve.

Starting at the top of the wall, we lightly misted the area with water from a hose, then pushed the pointing mortar into the clean, moist spaces between stones. I used a trowel or hawk to hold a big blob of mortar and pushed it in with a thinner trowel that just fit the spaces, until each joint was filled (photo facing page, left). The mortar we used was something like peanut butter in consistency, though not quite that sticky.

Once a section of the joint was full, I moved horizontally along the wall without stopping to pretty it up, leaving the mortar crude at this stage. It's much easier to dress later when mortar has lost its smeary quality and become crumbly. The less you manipulate the mortar, the better it will set. Just dragging a trowel across wet mortar brings too much lime and water to the surface, creating a slick, hard skin and leaving a sandier mix inside. You can mix half-set mortar back into a soft paste again (this is called retempering), but it will be weaker when it finally sets than if you had left it alone.

Adding extra water to a batch that's getting hard will greatly undermine its final strength. With bedding mud, I occasionally allow myself to do this, but I never do it when pointing. If you are pointing by yourself, keep your batches small (about nine shovels of sand), and mix them by hand in a wheelbarrow. Batches this size aren't worth the bother of a mixer.

Getting mud to the wall—Mobile mortar is a joy, so I work right out of the wheelbarrow when I can. On scaffolding, I transfer the mud to an old wheelbarrow pan or piece of plywood. By keeping the pan or board directly beneath the working area, I can catch most of the spills and reuse them. If the droppings get dirty, however, I use them to fill interior cavities in the wall.

Beginners are often frustrated by the amount

One advantage of bedding the stones first and pointing them later is that you don't need a strong bedding mix. The smears of mortar on the stonework clean up easily with a wire brush and water.

Starting at the top of the wall and working down, Kennedy holds the mortar on a standard mason's trowel (left) and packs the joints with a narrow pointing trowel. After the mortar has set long enough to become crumbly, the joints are dressed (right).

of mortar that falls off their trowels, but it just takes a while to get the feel of tools and mud. My best advice is to assume a lot will spill, and that it's okay. Position a catch board beneath you so you won't get upset by the waste, and wait to clean up flops that hang on the wall until later, when they become crumbly.

After using up a batch of pointing mortar, check the spot where you began. If the mortar is still smeary and sticks to the trowel, leave it alone. Timing can be tricky on a pointing job. It's predictable, but there are lots of variables. If you wait too long to dress the mud, it will be obnoxiously slow and hard to remove. On a warm dry day, the mix should be ready for dressing within an hour. On a cool, wet day, the first pointing of the day may not be ready until afternoon. A dry mix will set faster than a wet one; colder ingredients mean a slower set. Dry porous stone will absorb moisture from the mud and cause a fast, weak set. Direct sunlight, especially on black mortar, will also cause quick water loss and a weak product. Extra portland cement in the mortar gives a faster set; all-lime mixes are very slow.

If the mortar is crumbly, you can start dressing the joint. There are many schools of thought on finishing mortar joints. Some masons form a ccnvex protruding joint with a special trowel; some use various S-shaped metal jointers to cre-

ate a slick, hard surface. I prefer the look of a recessed joint. The stones become the dominant visual element this way.

Scrape off the lumps, and smooth out the irregularities until the joint is uniform. As shown in the photo above right, I use the tip of a 1-in. pointing trowel for this, though a stick will work, and I know one mason who uses a kitchen spoon. A brisk brushing with a small dry paintbrush or auto-parts brush will then tighten the joint by removing protruding grains of sand, and will bring a little extra lime and water to the surface of the joint.

By working in horizontal layers when pointing, you will avoid the dead look of a joint where one day's work ended and the next day's began. If you completed a vertical section on one day, then started next to it the following day, you would be able to see the junction (as with painting a house). If you can't finish a horizontal run in one day, end your efforts in jagged steps, for a less obvious blending the next day. Because every mason's dressing work looks a little different, it's also a good idea for a single mason to complete a given visual area.

Dressing joints usually takes me half as long as mixing and packing the mortar. If you underestimate this time, you're likely to point for six hours, start dressing down, and realize you won't get home for dinner. Then you're tempted

to finish the next morning (usually a horrible mistake, unless the mud was very wet and cold to begin with) or rush through the cleanup and get a messy-looking job. If you want to work an eight-hour day on pointing, don't do much more than five hours of packing the joints. Leave time to cover the cement and clean the tools.

Keep it wet—Mortar sets up. It doesn't (or shouldn't) dry. It is best if the mortar, as it is setting in the wall, is kept moist and between 50°F and 80°F for five days or so. (A month would be better, but let's be realistic.) This means no freezing, no heating, no rain, no drought. So if it isn't actually 60°F and drizzling lightly the whole time your mortar is setting, you must create those conditions.

Drape burlap over the finished areas and keep it moist by spraying it occasionally with a hose. Few masons bother with this step, but it is particularly important with these primitive lime mortars. My attitude is to be loose about the bedding mix but persnickety about the pointing job, since it has so much more to do with the looks and longevity of the stonework. □

Stephen Kennedy lives in Orrtanna, Pa. His book Practical Stonemasonry Made Easy *is available from TAB Books Inc. (P.O. Box 40, Blue Ridge Summit, Pa. 17214; $16.95).*

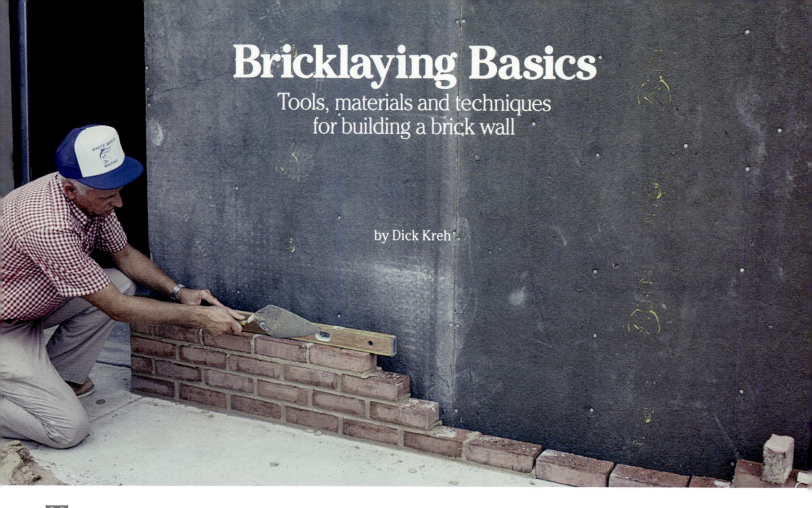

Bricklaying Basics
Tools, materials and techniques
for building a brick wall

by Dick Kreh

The art of laying bricks in mortar to form a wall dates back thousands of years, and the tools haven't changed much over the centuries. Whether your project is simple or complex, laying bricks can be one of the most creative and satisfying of all crafts. Bricklaying is not complicated; the key to good work is consistency, accuracy and repetition. This requires strict attention to details, such as the placement of your fingers when holding the trowel. With practice you develop a feel for the work.

You probably already have many of the tools you'll need—a steel measuring tape, a metal square, a ball of nylon line, a brush, a chalkbox and safety goggles or glasses. More specialized bricklaying equipment is described in the sidebar on pp. 74-75. For mixing mortar, you will need a wheelbarrow, a mixing box (for large batches), a water bucket, a 24-in. sq. mortarboard or pan, a mortar hoe (like a garden hoe, but it has two holes in the blade) and a hose with a spray nozzle. If a lot of mortar is to be mixed, utility drum mixers can be rented.

Mortar—A wall can be no stronger than the mortar it is built with. If the mortar is too weak, the wall will fail. If the mortar is too rich, it will be sticky and hard to handle with the trowel, and will be brittle when it hardens.

There are several kinds of mortar, but they are basically all composed of portland cement, lime, sand and certain additives that make them more workable. The simplest to use is prepackaged

Dick Kreh is an author, masonry consultant and building-trades teacher. He makes his home in Frederick, Md.

mortar, which contains all the dry ingredients in the correct proportion; you add the water. Prepackaged mortar is fine for small jobs, but on a project of any size it is expensive. Mortar is mostly sand, and sand is inexpensive.

The most popular mortar for general use is made with masonry cement. It is sold in 70-lb. bags (with lime and certain additives included), and you add the sand and water when mixing. A good proportion to mix is 1 shovelful (part) of masonry cement to 3 shovelfuls (parts) of sand, with enough water to blend it to the desired stiffness. If you mix an entire bag at once (called a batch), use 1 bag of masonry cement to 18 shovelfuls of sand. Masonry-cement mortar is the most economical and has excellent handling properties.

With portland cement/lime mortar, you can adjust the strength of the mix by varying the ingredients. Building-supply stores sell portland cement in 94-lb. bags. The lime comes in 50-lb. bags and must be labeled "Hydrated" or "Mason's Lime," which means that it has been treated with water. This is stamped on the bag. Other types of lime are used for agricultural purposes.

The portland cement/lime mortar mix I use is 1 part portland cement to 1 part hydrated lime to 6 parts sand and water. This is known as Type N mortar and is the standard mix for brickwork. After curing 28 days, this mix will sustain pressures of 750 lb. per sq. in., which is more than ample for most brick masonry.

A word of caution about sand. Buy your sand from a regular building supplier and ask if it is washed. This is important, because with washed sand most of the loam and silt will have been removed. Silt or loam in the sand will prevent

the mortar mix from blending together properly; the resulting mortar will be weak and defective.

Store cementitious materials in a dry place and off the ground so moisture won't get in. If moisture has penetrated into the dry cement and hard lumps have formed, throw it away. If your project is large, it is better to make several trips to the supply house for cement than to keep a lot of it for six months or a year—mortar stored too long tends to lose its strength.

When measuring, proportion the materials carefully, and mix all the dry materials together well before adding clean water. Don't mix any more mortar than you can use in an hour. If the mortar starts to stiffen in the wheelbarrow, you can add enough water to temper (loosen) it, once—this should be only enough water to make it workable, not runny or thin.

Estimating materials—Over the years I have developed some rules of thumb for figuring how many bricks and how much cement, sand and lime to order. These may be of some help to you when you estimate your project.

Figure on using seven standard bricks (a standard brick is considered to be 8 in. long, 3½ in. wide and 2¼ in. high) for every single thickness of 1 sq. ft. of wall area. This allows a small amount for waste, so don't add any more to it for cracked or broken bricks.

One bag of masonry-cement mortar will lay about 125 bricks. One bag of portland cement and one bag of mason's hydrated lime mixed with 42 shovels of sand will lay about 300 bricks. One ton of sand will make enough mortar to lay 1,000 bricks. If you store it on the ground, allow about 10% for waste.

Layout—Since bricks vary somewhat in length, you can't just step off the bond (the pattern for layout that will be followed throughout the wall) with a tape measure. You have to lay out the first course brick by brick to a chalkline. This is done most effectively by spacing the bricks ⅜ in. apart (the width of the head joint, or space between the ends of adjacent bricks) without using mortar—this is known as *dry bonding*. A quick way to lay out the head joints is to place the end of your little finger between the bricks. When the first course is aligned, if there is a small gap at the end, either open up all the joints or place a cut brick in the center of the wall or under a door or window. Try to have as few cut bricks as possible. Be sure to lay the face side of the brick out to the front of the wall. (The face side is usually the straightest side, and it always matches the ends of the brick in color.) Leave the bricks in position until the mortar is ready to spread. Then pick them up only two at a time so you don't lose the pattern or spacing. Only the first course is laid out dry. Subsequent courses will be laid following the pattern established by the first course.

Cutting a brick—Bricks can be cut by using either a brick set chisel or a brick hammer. Always check each brick to be cut. If it has a crack in it, discard it because it won't break true. To make an accurate cut, first mark the edge to be cut, and then place the flat side of the blade facing the finished cut. This will ensure a neat, accurate cut. Be careful not to place the fleshy part of your hand against the head of the chisel—the hammer knows no mercy if misdirected. It's a good idea to wear a pair of gloves to prevent accidentally cutting or pinching your fingers.

Most novices have trouble cutting bricks with the brick hammer, but as with so many things, practice makes perfect. Mark the cut with a pencil first, then score across the face of the brick with light pecking blows. Don't try to break the brick yet; the object here is to weaken the brick along the line.

Next, repeat the scoring of the brick across its widest side. It should break cleanly. If not, return to the face side, and repeat the scoring until the brick breaks at that point. Now trim the cut edge with the end of the blade to remove any protruding edges.

Troweling techniques—Cutting and spreading mortar are important skills to master. Of the many different methods used, I feel the easiest is the "cupping" method. It is accomplished in a series of steps or movements.

Start by holding the trowel with the thumb just a little over the end of the handle, near the blade (photo top left). This will provide you with good leverage and balance. It also keeps you from dipping your thumb in the mortar. Your fingers should be wrapped firmly around the handle, but not gripping tightly.

Cut the mortar away from the main pile with a downward slicing motion, pulling it toward the edge of the board (photo top left). Shape the mortar on both sides with the blade of the trowel until it is about the same width and length as

Troweling the mortar. With a downward slicing motion (top left), cut some mortar away from the pile in the middle of the mortar board, then push it into a rough box shape the size of the trowel blade (top right). Now slide the trowel under the mortar so that it fills the blade evenly (middle photo), snap your wrist down to set the mortar on the blade, and spread the first bed joint (above).

the trowel blade. One or two motions will do this (photo top right). Then, with a smooth forward motion, slide the trowel under the mortar (middle photo above). Lift the trowel, and at the same time, snap your wrist slightly down to set the mortar on the blade. This keeps the mortar on the blade for the spreading operation.

The mortar is now ready to spread to form the first bed joint (the joint between courses). Move the arm in a sweeping, spreading motion,

keeping the point of the trowel in a fairly straight line and turning the blade sideways at the same time. This is done with a flowing motion for best results. The mortar should roll off of the trowel blade, following the point, in a straight line (photo above). With practice, you will get the same depth of mortar, ⅜ in., for the entire spread. Practice until you can spread about 16 in. at one time.

Mortar spread for the first course of brick on

Maintaining plumb and level. Check the first brick for vertical alignment with the modular scale rule (top), then level its top (middle photo) and plumb its face (above). Repeating this procedure at the opposite end of the wall gives two points to pull a line from.

A line block hooked over one end of the wall is held in position by a line block at the other end. A dry brick set on top of the line sets it in the plane of the face of the wall.

the base or footing should not be furrowed (indented) with the trowel. It should be solid to prevent water from leaking through.

The first course—Establishing and maintaining plumb and level are critical in any brickwork. Beginning at one end of the wall, lay the first brick, pressing it into the mortar bed with your hand as level and plumb as possible, by eye. Then check with the modular scale rule to see that the top of the brick aligns with 6 on the scale side of the rule (photo top left). Now level the brick from the measured end until the bubble is between the lines on the vial (photo second from top). Then plumb the face of the brick, being careful not to move it out of alignment with the layout line (photo third from top).

Repeat the procedure at the opposite end of the wall so that you have two points to pull a line from. Do this by attaching the line and block over the end of the brick on one end and pulling the line tightly over the end of the brick on the opposite end. Wrap the line around the brick so it won't slip loose. Block (push) the line out to the exterior face of the brick by laying another dry brick on top of it, as shown in the photo bottom left.

With the line in place, pick up two bricks at a time from the layout and spread mortar there. Butter a mortar head joint on one end of each brick as it is laid. This is done by picking up some mortar on the point of the trowel, setting it with a jerk of the wrist on the trowel, and then swiping it on the end of the brick. With practice, one swipe will do the job.

As the bricks are laid in the mortar bed joint, be sure to keep them about 1/16 in. back from the line. This will prevent the wall from being pushed out of alignment by bricks riding against the line. If you lay the bricks too far back from the line, the wall will bow in the other direction. Bricks against the line are called "hard" and bricks back too far are called "slack," and neither is acceptable. After a little practice, you will get the hang of this.

Lay the bricks from each end of the wall to the center. If the spacing doesn't work out perfectly, either open up the head joints a little or you may have to cut a brick. The general rule for cuts in the center of the wall is not to make bricks smaller than 6 in. Sometimes you need to cut two bricks to make the wall look right. Locating the cut brick under a window or door makes it less conspicuous.

If a brick wall is going to leak, it will usually be at the spot where the last brick was laid. Therefore, to prevent this, butter each end of the bricks already in place and each end of the brick to be laid. This may seem time-consuming, but will pay off by preventing any future problems. This last brick is known in the trade as the "closure" brick.

Building the corner or lead—Once the first course has been laid out in mortar, the ends or corners (also called leads) should be built up about nine courses first, instead of trying to lay each course individually. Build the ends of the wall to the desired height, and then the line can be attached with line blocks and filled in be-

Tools and equipment

You won't need much equipment to lay bricks, so it makes sense to buy brand-name, good-quality tools, because they are better balanced and are a pleasure to handle. Bricklaying tools can be obtained at good hardware stores, building-materials suppliers or from mail-order catalogs. One tool company that I have dealt with over the years with complete satisfaction is Masonry Specialty Co. (4430 Gibsonia Rd., Gibsonia, Pa. 15044). Two others are Goldblatt Tool Co. (511 Osage, P.O. Box 2334, Kansas City, Kan. 66110) and Marshalltown Trowel Co. (Box 738, Marshalltown, Iowa 50158). When you need costly equipment or tools, the best solution is to rent them locally. Check the Yellow Pages in your phone book under Rental Service Stores and Yards.

I consider the following tools essential for brick masonry work.

Brick trowels—Trowels for applying mortar to bricks range in length from 10 in. to 12 in. Brick trowels 11 in. long work best. I like a narrow blade, which is called a narrow London pattern, with either a wood or plastic handle. A flexible blade is a must for spreading and applying mortar joints on the edges of the bricks. You can test a blade's flexibility by bending the point against a flat surface. It should flex about 1 in. to be effective. Two very good trowels are made by W. Rose Inc. (P.O. Box 66, Sharon Hill, Pa. 19079) and Marshalltown. However, you may want to pick your trowel out in the hardware store in order to test the flex of the blade.

Brick hammers—Brick hammers are used mainly for cutting bricks. Select a well-balanced brick hammer that weighs about 18 oz. Holding it by the handle, see if it feels comfortable and balanced. The handle may be wood, metal or fiberglass. I like a wood handle. There are many good brick hammers on the market. A few brands that I recommend are Stanley, True Temper and Estwing.

Levels—Of all the tools that you will need, the level is the most delicate and probably the most expensive. It is used to keep your work level and plumb. Levels may be metal, fiberglass or wood. I prefer a wooden level with alcohol bubbles because changes in temperature will not cause the bubble to shrink or expand. Levels come in various lengths. If you are going to get only one, buy a 48-in. level, which fits most situations. If you are going to buy two, also buy a 24-in. level. Any brand name will do, but be sure that the level is true before leaving the store. Do this by holding it in a plumb position alongside another level. The bubbles should read the same, top and bottom. Reverse the edges and check again. Now try the test holding the levels horizontal. Two good levels I like are the American (Macklanburg-Duncan, P.O. Box 25188, Oklahoma City, Okla. 73125) and Exact I Beam (Hyde Manufacturing Co., 54 Eastford Rd., Southbridge, Mass. 01550).

Tool photo, facing page: Michele Russell

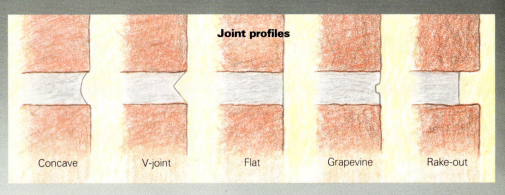

Joint profiles

Concave   V-joint   Flat   Grapevine   Rake-out

Chisels—Chisels are useful for cutting bricks to length, and for removing hardened mortar. For most work, two basic chisels will suffice: a flat broad chisel with about a 3½-in. blade (called a brick set) and a standard mason's cutting chisel with a 1½-in. blade and a long narrow handle.

Jointers—Jointers, or strikers, are used to form, seal or finish the surface of mortar joints on brick, block or stone work. They can add a sense of design, depth or texture to the wall. Jointers should be made of good-quality steel and be reasonably priced. The sled-runner one I use costs about $8.

Jointers for brickwork come in five common shapes (drawing, above right). Convex jointers, the most popular, form a half-rounded, indented profile (photo above left). V-jointers give an angled indented finish that

looks great on rough-textured brick or block work. Slickers form a flat, smooth finish and are used for paving steps or repointing old mortar joints. The grapevine profile has a raised bead of steel on its edge that forms a wavy line when passed through the center of the mortar joint. It is very popular for early American brickwork. The rake-out forms a neat raked mortar joint that looks good with sand-finish or rough-textured bricks. You can get one mounted on two skate wheels to keep you from skinning your knuckles. The raking is accomplished by a nail in the center; the depth of the rake is adjusted by a thumbscrew. Oiling around the wheels will keep the tool in good condition.

Mason's rules—Two types of masonry scale rules can be used for brickwork: the modular rule and the course counter rule. The modular rule is based on a 4-in. grid (the basic module for manufactured construction materials). On one side of the rule is the standard 72 in.; the other side has scales for bricks of various heights. The standard mortar joint of ⅜ in. is used for all sizes of bricks. You will probably use scale #6 the most, which means that six courses of standard bricks, including the mortar bed joints, will equal 16 in. in height. Scale #2 is for concrete blocks (8-in. increments). The other scales are used less often. The course counter rule is designed for

all standard-height bricks but allows the user to vary the thickness of the mortar bed joints. For most work, the modular rule will suffice.

Line blocks and pins—Line blocks are used to attach a line to the corners of a wall as a guide for laying bricks between them. They are paired wood or plastic blocks with a slot scored on one side and an end for passing the line through. They are held on the wall by the tension of the line pulled tightly between the corners. Building-supply houses often give them away as promotions.

Like line blocks, steel line pins can also hold the string line. They are driven in the mortar head joints, and then the line is wrapped around and pulled tightly from one end of the wall to the other. Line pins are also available on request from most building-supply houses. —*D. K.*

Tools for brick masonry

Pointing trowel

Narrow London-pattern trowel

Modular rule

Line blocks

Line pin

Brick hammer

Brick set

Standard mason's chisel

Convex jointer

V-jointer

Slicker

Grapevine jointer

Skate-wheel rake-out jointer

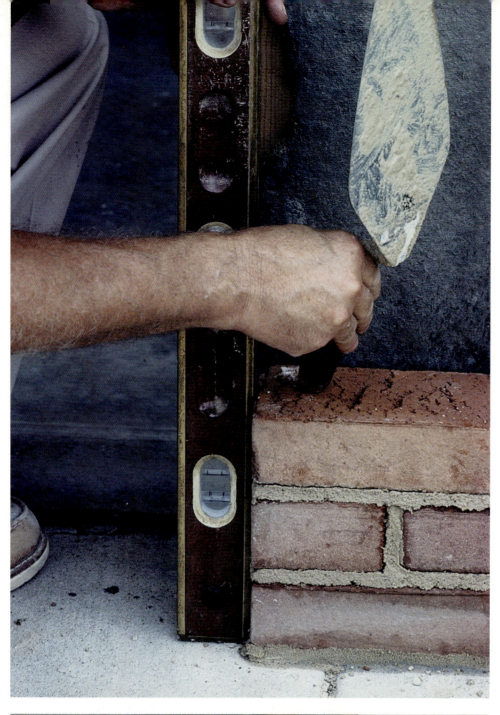

tween. Building the leads this way holds the ends of the line firmly in place.

As on the first course, maintaining plumb and level is a must. After laying each course, check the height with the scale rule, and then level out from that point (photo p. 72). Next, plumb the ends and front of each end of the course with the level (photo top left).

Now, holding the level horizontally against the face at the top edge, line up the course from one plumb point to the other. Whatever you do, don't change the original plumb points, but adjust the bricks between either in or out to meet the edge of the level. This is known as ''straightedging'' the course (photo bottom left).

Laying bricks to the line—The heart of bricklaying is laying bricks to a line. It is by far the fastest and most accurate method of building a brick wall. The mortar has to be handled differently here than when mortaring the first course. First, be careful with the trowel blade when spreading mortar in back of the line; if you hit it, the sharp blade may cut the line. Also, if you are working with another person, and the line is constantly moving, it will be very difficult to see exactly where the line is.

When laying one course of brick on top of another, you furrow the center of the mortar joint with the point of the trowel by hopping it up and down with a wrist action. Try not to punch all the way to the bottom of the brick below, but just make a good indentation. This is done for two reasons: it evens the thickness of the mortar on both sides, and the furrow creates a suction between the brick and the mortar. It is also much easier to press the brick into position to the line (photo facing page, top left).

After spreading the bed joint, cut or trim off the excess mortar from the front edge by holding the trowel on a slight angle (middle photo at left, facing page) to reduce smearing the face of the bricks below and to keep from tearing the mortar loose from the front of the joint. Return this mortar to your mortar board for future use.

Now lay the brick in the mortar bed, holding your thumb on the top edge, letting the line slip under it. Squeeze the brick back against the one previously laid to form a full mortar joint (bottom left photo, facing page). Don't drop the brick in the mortar, but ease it down so it won't sink too deep. This is one of the most common problems for a beginning bricklayer.

When you press the brick into position to the line, let the palm of your hand rest half over the brick previously laid and half over the one you are laying. You will be able to feel when the brick is laid approximately level. At the same time, cut the protruding mortar off on an angle with the trowel (photo facing page, top right). Be sure that you maintain the 1/16-in. spacing from the line.

Apply the mortar cut off to the end of the brick, to serve as the head joint for the next brick to be laid (photo facing page, bottom right). This is done with a swiping action under the line against the end of the brick. Apply mortar to the front and back edge of the brick for a more waterproof joint. When laying the closure brick, always double-joint the ends.

Plumbing and straightening. **As the wall gets higher, bricks are continually checked for alignment (top). Adjustments are made by tapping with a trowel (above).**

Buttering the bricks. On all bed joints but the first, the mortar is furrowed with the point of the trowel to make it easier to press the bricks into position and to create suction between the brick and the mortar (top left). When the bed joint is spread, cut off the excess mortar from the front edge (middle left), then lay the brick and squeeze it back against its neighbor to form the joint (left). With the brick in place, again trim the bed joint (above), and apply this mortar to the end of the brick, so it can serve as the next head joint (right).

Tooling the joints—Neat, smooth mortar joints look good, are more waterproof, and improve the structural integrity of the wall. The mortar joints can be tooled (struck) with various jointing tools. In the top left photo on p. 87, I am using the convex jointer to form a concave joint (half-round) finish. I prefer the long sled-runner type over the small pocket-size striker, as it will form a much straighter joint.

Head joints should be tooled first to ensure a continuous, clean bed joint and better appearance. Always tool the joints as soon as they are "thumbprint" hard. If they get too hard, they may turn black when they come in contact with the metal tool. Press your thumb in the mortar joint. If it leaves an impression, it's time to strike the joint. Usually, the joints are ready to be tooled about ten minutes after being laid. But the dryness of the bricks and weather conditions are the determining factors. If you wait too long, the joint will require too much pointing, and the finish will not be smooth. Fill in any holes in the joint as the striking progresses.

After striking, trim off any excess mortar and smooth the edge of the bed joint at the bottom of the wall with the point of the trowel. When the mortar joints have dried (set) enough so they won't smear, brush them lightly with a soft

brush to remove any remaining particles of mortar. At this time, if necessary, restrike the joints to effect a crisp, clean appearance.

There are going to be times when you will need to apply a mortar joint to the long side of a brick, such as when laying a cap on top of a wall or when building brick steps. This is done with two distinct movements. Pick up mortar from the board and set it on the trowel blade with a slight snap of the wrist. Then, holding your fingers under the brick, swipe the mortar down across the side, sticking it to the brick.

Pick up another trowel of mortar the same way, and swipe it on the bottom half of the side. If done correctly, a full mortar cross joint will result, and it will be waterproof. These motions should be done with some force so the mortar will stick to the brick. As you may surmise, some practice is required to master this.

Cleaning brickwork—After tooling the joints, let the mortar cure at least a week before cleaning the wall. You can use the old standby, muriatic acid and water, or one of the many masonry cleaners now on the market. If you use muriatic acid (which is safe if used according to directions), mix 1 part muriatic acid to 10 parts water in a plastic or rubber pail. Always put the water

in first and then add the acid. Mix the solution outside to minimize inhaling the fumes. Muriatic acid is available from your local building-supply dealer in quarts or gallons, and is inexpensive. Sure Klean 600 Detergent (ProSoCo, Inc., 755 Minnesota Ave., Kansas City, Kan. 66117) is another good masonry cleaner. It, too, is available at most building-supply houses.

Whatever cleaner you use, always follow the directions on the container. If you do get any cleaner in your eyes or on your skin, flush immediately with plenty of water. Wear eye protection and rubber gloves while you work.

After covering nearby shrubbery or metal window frames in the area to be cleaned, begin by rubbing off any loose particles of mortar with a piece of wood or paddle. Next, soak the wall from top to bottom with a hose. Then scrub the cleaning solution on the brickwork from top to bottom with a good stiff bristle brush, like the ones used in farm dairies. Work in an area about 12 ft. square at a time. Stubborn spots or splatters of mortar can be removed by rubbing with a piece of brick and then scrubbing again. Be sure to flush off the cleaned areas top to bottom with water under pressure from a garden hose with a spray nozzle until the water runs clear. This also neutralizes the cleaning solution. □

Putting Down a Brick Floor

The mason's craft is easier when the bricks are laid on a horizontal surface

by Bob Syvanen

Brick floors are experiencing a revival. The material used only for patios or walkways 10 years ago is moving inside the house. There are good reasons for this. The color, texture and pattern that brick brings to a room can't be duplicated with vinyl or wood. And brick is a logical choice for a passive-solar building, because it increases the thermal mass while offering a more finished look than a concrete slab.

Laying a brick floor sounds risky if you're not skilled with a trowel and a brick hammer. But after watching master mason John Hilley lay the floor in a house I'm building in Brewster, Mass., I'm convinced that it's not too difficult. As with any job, if you know the tricks, the battle is half over.

A brick floor laid with mortar is different from a wall or a patio. First, you can ignore plumb and just concentrate on level. Second, you don't have to contend with the shifting, subsiding backyard quagmire that is the substrate for most patios. A brick floor should be laid on a concrete slab, which is flat and solid; or on a wood floor that has been beefed up to take the extra weight of the bricks and stiffened so that its flexibility won't crack the mortar joints.

Brick—If you think that brick is brick, and that your only decision will be how many to buy, you'll change your mind when you hit your first masonry-supply yard. Bricks come in many styles, colors and prices. They run anywhere from 30¢ each to 50¢ each or more. If your floor extends out into the weather, use paver bricks. The surfaces of pavers are sealed, so the bricks won't spall when it freezes. If your floor is inside, anything that strikes your fancy will do.

Used bricks make a very nice floor but they're getting scarce, and as a result, expensive. You never know exactly how much waste you'll get when you scale the old mortar off used brick, but plan on buying at least 3,000 used bricks to get 2,000 usable ones. For new 4x8 bricks, you should figure 4½ bricks per square foot of floor and then add at least 5% for waste.

Since a standard brick weighs about 4½ lb. and you're going to need a lot of them, you'll want them delivered to a convenient place. Most yards bring the bricks on pallets to your job site, and if the delivery truck is equipped with a hydraulic arm, you're even better off. The arm lifts the wood pallets off the truck and sets them down anywhere you say. A skilled driver can just about put the bricks in your back pocket.

Use brick tongs for hauling the bricks from the pallets. Brick tongs are simple tubular-steel contraptions that will carry between 6 and 10 bricks at a time by holding them in compression. At about $16, using tongs beats weaving around the site with a pile of bricks stacked up against your forearm.

Layout—Although there are many patterns or bonds that bricks can be laid in, the *running bond* is still the easiest and one of the most attractive. The joints between the bricks in each course are offset from the joints in adjacent courses by a half brick. This means that each course begins with either a half brick or a whole brick. After sweeping the slab absolutely clean, determine which direction the brick courses will run. You can then begin the rough layout of the floor with a tape measure.

To avoid having to cut a course of narrow bricks at the end of the room, you may need to adjust the width of the joints. Laying out to a full course at the end of the room is time well spent. Do your figuring on paper by adding an ideal joint width, for example, ¼ in., to the width of your brick and dividing the total length of the room by this sum. Then confirm your calculation by *dry coursing*—laying a full row of bricks along the length of the room without mortar to test-fit the layout. Pick carefully for representative bricks, since they can differ considerably in length and width. Remember that thin joints look better than fat ones when you're making adjustments between courses; and that you'll get another inch at each wall to play with because the wall finish and baseboards that will be installed later will cover that much more of the floor.

Once your dry coursing has been adjusted so that the joints are even, nail 1x or 2x layout boards to the wall studs along each side of the room so that their tops are even with the top of the finished brick floor (drawing, below). Mark the leading edge of each course on the board, and drive a nail into the top of the board at each of these marks for a string line. After laying a course, move the string forward one nail on each side of the room for the next course. This line represents both the finished height of the floor and the leading edge of the course, leaving very little for you to eyeball.

A layout board can also be cantilevered off the top of bricks that have already been laid. This setup works well when an exterior wall takes a jog, making the room narrower, and ending a run of layout boards. Course lines are marked on the end of the board that projects out to the unlaid part of the floor, and the course string is attached so it rides on the bottom edge of the board. This maintains the same finished floor height as layout boards held flush with the top of the bricks.

You should dry-lay a test course along the width of the room, too. You will be starting with a half brick or a whole brick on one end, and adjusting the joint width between the ends of the bricks to determine the length of the brick on the other end. It won't always work out to half and whole bricks, but the less cutting you have to do, the easier the job will be. Don't end a course with a very short brick.

For anything wider than a closet or a hallway, use *control bonds* to make sure that the bricks are being spaced uniformly. These are bricks whose ends are laid to a string as a reference every 6 ft. or so (about 10 bricks) within each course. This in effect breaks a long course up into several small courses.

The control bonds and end bricks are the

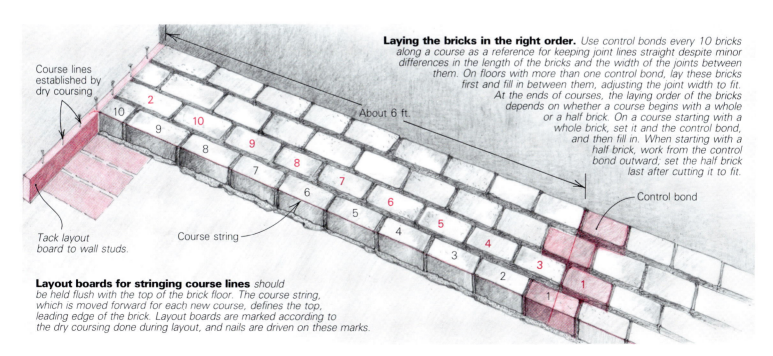

Course lines established by dry coursing

Tack layout board to wall studs.

Course string

Layout boards for stringing course lines *should be held flush with the top of the brick floor. The course string, which is moved forward for each new course, defines the top, leading edge of the brick. Layout boards are marked according to the dry coursing done during layout, and nails are driven on these marks.*

About 6 ft.

Laying the bricks in the right order. *Use control bonds every 10 bricks along a course as a reference for keeping joint lines straight despite minor differences in the length of the bricks and the width of the joints between them. On floors with more than one control bond, lay these bricks first and fill in between them, adjusting the joint width to fit. At the ends of courses, the laying order of the bricks depends on whether a course begins with a whole or a half brick. On a course starting with a whole brick, set it and the control bond, and then fill in. When starting with a half brick, work from the control bond outward; set the half brick last after cutting it to fit.*

Control bond

first bricks to be laid in each course. The rest are filled in to fit. This way, the joints of every other course at the control bond will form a straight line, and the cut bricks at the end of courses, as well as the width of the mortar joints at the ends of the bricks, can be kept fairly consistent from course to course. Try to place control bonds in highly visible spots such as stairways and entries, and string them just as you did the course lines.

Stock the floor once you've completed your layout. This way you can stretch strings and get to know the peculiarities of the room and the slab you'll be working on before you begin littering it with bricks. Using brick tongs, distribute the bricks so that they will be within easy reach when you begin to work. Keep in mind that several layers of bricks on a pallet, or even the whole thing, can be a very different color from the rest of the load. Mix these colors and tones as you stock them on the floor so that your floor doesn't end up with big patches of only dark or light bricks.

Tools—A brick floor is laid with standard mason's tools (photo below). In addition to a 4-ft. level, a tape measure and nylon string, you'll need a brick hammer to break the brick to length at the end of a course. It has a square, flat head and a long, flat chisel peen on the other end, and is made of tempered steel. Brick hammers come in various weights. They have either steel, fiberglass or wooden handles, and cost about $15.

Brick trowels also cost about $15, and are made with wood or plastic grips. There are two basic shapes: the London and the Philadelphia. Most brick masons prefer the London pattern, an elongated diamond shape with its heel farther forward on the trowel than the Philadelphia. London patterns come with either a narrow or a wide heel. The narrow heel is fine for brick since less mortar needs to be carried by the trowel than for stone or concrete blocks.

You'll need a jointer for smoothing and shaping the mortar between the bricks. This tool looks like an elongated steel S, and is gripped in the middle. Each end of the tool has a different profile.

Tools of the trade. **Brick tongs, top, make stocking the floor with bricks much easier. The trowel, jointer and brick hammer below it are the basic tools used to lay the floor.**

Mortar—The mortar between the bricks in your floor makes it permanent, and provides a visual relief from the brick itself. In this case, it is a mixture of masonry cement, sand and water. Use a shovel and buckets to proportion the ingredients for the mortar. Mix your mortar with a hoe in a mortar box or a wheelbarrow. An easier way to mix is with a mechanical cement mixer. Do your mixing outside where you are free to hose out your mixer at the end of the day, but keep the fresh mortar out of the hot sun.

The amount you'll need depends on two things: how much bedding is required for the slab you are working on, and the thickness of the joints between bricks. If your slab is flat and level, the bed of mortar under the bricks will be fairly uniform. A good bed is ⅜ in. thick, and no less than ¼ in. However, a serious hog, or hump, in the floor can double the amount of mortar, because you will need to bring the bricks for the rest of the floor up to this level with a much thicker bedding.

The width of the joints between bricks is the other factor that affects how much masonry cement and sand to order. Joints look bigger than they really are because bricks are molded and don't have hard crisp edges. A ¼-in. joint after finishing will look ⅜ in. wide, which is a nice size. Figure on one bag of masonry cement for every 100 bricks if your slab is uniform and your joints are ¼ in. wide. You will need 1½ cu. ft. to 2 cu. ft. of sand per bag of masonry cement.

The consistency of your mortar will have a lot to do with your success in laying bricks. Mortar should be firm enough to support a brick as bedding, yet soft enough to compress easily at the joints. A soupy mix will lay down, or self-level, in the wheelbarrow. A mix that's too stiff will support itself even when it's stirred into the shape of a breaking wave. A workable mix is the consistency of whipped cream, and the secret to mixing it that way is adding water to the dry ingredients in small amounts, and lots of practice.

A good mason mixes only the amount of mortar that can be made with one bag of masonry cement at one time. With this amount, the consistency is easy to control. A batch will last about three or four hours before setting up. Temper the mortar every 15 minutes by working it briskly with a hoe or shovel and adding a little water if necessary. The books say not to add water, but all masons do. Whether the mud is mixed with a mechanical mixer or in a tray with a mortar hoe, it's best loaded into a wheelbarrow because it is convenient to move around the work area, and it's easy to scoop out of.

Laying the bricks—You are ready to begin laying bricks when the mortar is mixed and the control-bond strings and the first-course string are stretched. For courses that begin with a full brick, lay the control bonds first, the full end brick second, and then fill in between, as shown in the drawing on the previous page. Start from the control bond and work outward. If the course starts with a half

brick, lay the control bond first, and then work from this brick out toward the ends, cutting the bricks on either end to fit. A little gain or loss that accrues as a result of the inconsistent length of the bricks can be offset by adjusting the size of the last brick.

You cut the last brick in a course with the mason's hammer. Hold the brick in your hand, and hit it sharply once or twice directly over your palm. This usually does it, but some bricks need a shot on both sides. Sometimes they break where you want, and other times you end up with a handful of brick shards. You'll get better with practice, but until you do, order enough extra brick so that each blow isn't critical. If the brick fractures at an angle, set it down on a hard surface and use small, chipping strokes with the hammer to straighten out the line of cut.

Another tool for cutting brick that requires a less practiced stroke is the brick set. This wide chisel is placed on the brick where you want it to fracture, and struck with a hammer (for more on breaking brick, see pp. 73, 85, 105). A brick set will cost you about $6. Plan to waste a few bricks using this tool as well.

You can tell a journeyman from an inexperienced mason just by watching his trowel hand as he picks up a load of mortar and places it. There is a familiarity with the material that is unmistakable. First, using the back of his trowel in the surface of the mortar, he will stroke away from his body. This creates a mound of mortar in the wheelbarrow. With the face of the trowel he scoops a load of mortar in an upward motion, then drops his trowel arm abruptly, ridding the trowel of excess mortar. What remains won't slide off the trowel even if it's turned upside down. The technique is stroke away, scoop up, and settle.

The first trowel of mortar should be placed on the floor for bedding. Thrown is more accurate, though this too takes a bit of practice. Then choose a brick. If your bricks are water struck on one face, this face should be laid up because its slightly glossy surface is less porous and will wear better. With brick styles that are water struck on all sides, or not at all, just choose the best face.

With the brick in one hand and the trowel in the other, pick up a thin line of mortar using the same stroke as before, and wipe it on the edge of the brick. As you get more experienced, it will almost look like you're throwing the mortar on. Unless you overload the brick, the mortar shouldn't slip off while you are handling it. Trial and error will tell you how much mortar to use. Finish buttering the brick by loading up the end that will butt the previously laid brick.

To set the brick in place correctly, all of its joints need to be in compression. This is one of the secrets of good brickwork, and the way to accomplish this is the *shove joint*. The brick should be held out from its ultimate resting place and square to it. As you begin to bed it down, the brick should sit slightly above the string line on the mound of mortar underneath it (photos facing page). Using your thumb, push the side of the brick toward the

The shove joint. **Above, mason John Hilley beds a brick in a running bond. It is buttered on one side and one end with mortar, and will sit well above the course string on the bedding mortar until it is pressed and tapped down. This string indicates the top of the floor, and the leading edge of each course. It is moved ahead one nail on the layout boards at each side of the room after a course is completed. The brick is leveled to the string by a gentle tapping with the butt of the trowel (above right), while pressure is applied with the left hand. This compression is the key to a tight, permanent brick floor. The lines of mortar between bricks are jointed, right, when they reach the consistency of putty, using excess mortar on the surface of the brick. The mortar is fed into the joints and compressed with a jointer. This is done twice on each course to produce smooth and compact mortar joints.**

previously laid bricks and use the butt of the trowel to tap the brick down until its face is just below the string. If the brick is too low anywhere, lift it out and load the bed with more mortar. Then press the brick into place again. Also use the trowel handle to tap the end of the brick until the width of the joint between it and the previously laid brick is correct. During all of this you should be pushing the side of the brick with your thumb. This pressure keeps the brick from settling unevenly, and keeps the joints in compression. Don't worry if the mortar doesn't rise all the way to the surface in every spot along the joints. The holes will get filled in later.

Scrape away any excess mortar with the edge of your trowel. This mortar can be allowed to dry a little and be used for jointing. Try not to smear the mortar on the surface of the bricks, but if you do, don't worry. It can be washed off later after it sets up.

Jointing—The brick joints are ready to be filled and jointed when the mortar between the bricks has the consistency of putty. Test it by pressing the mortar with your finger. You can use the scrapings on the surface of the brick to fill the joints if they hold together when you squeeze them in your hand. They shouldn't be soft like fresh mortar, or they won't compress when jointed. Dried or crumbly mortar shouldn't be used either.

The joints should be filled and compressed in two stages. For the first fill, feed the mortar into the joint and press it with the jointer

(photo above right). Work your way along the course, filling and pressing. Then go back to the beginning of the course for a second fill and a final jointing. Jointing not only increases the strength and durability of the mortar joints, but it also gives the floor a smoother, more uniform appearance. Do the final jointing with smooth, level strokes. Use your whole arm, not just your wrist.

Cleanup—Let the mortar set for a week before cleaning. You can wait longer, but the brick will get harder to clean. First, scrape off heavy spots of mortar with the chisel peen of a mason's hammer. To remove any mortar smeared on the surface of the bricks, you will need to use a 20% solution of muriatic acid. Mix the acid in a bucket by adding one part acid to four parts water. Always add the acid to the water, and make sure that you are wearing eye protection and heavy-duty, acid-resistant rubber gloves. You will also need running water from a garden hose close by. Open as many doors and windows as you can for ventilation. Muriatic acid is nasty stuff—treat it with respect and caution.

The acid will soften the mortar spots, enabling you to scrub them off the surface of the bricks. For this you will need a short, stiff-bristled brush for hand-scrubbing, and a 9-in. stiff-bristled brush on a long handle. Before you begin, flood the whole floor with water until the bricks are saturated. This keeps the acid action on the surface where it can be controlled, and then washed off as soon as it

has penetrated the mortar smears. The bricks will look shiny when they are saturated, dull when they are beginning to dry out. Keep the bricks saturated during the entire process so the acid won't burn the mortar. Get a helper to operate the water hose.

As soon as you mix the acid solution, test its action in some hidden spot—under the stairs or behind the chimney. The joints will have a soapy appearance as long as muriatic acid is present. It's best to do this job systematically and work small areas. Dip the brush into the acid solution and then begin scrubbing the saturated bricks. As soon as you have covered a few square feet, hose it down until all the foaming stops. Then hose it a bit more, saturate the next area, and move on with the bucket and brush. After the entire floor has been scrubbed and rinsed, hose it down one more time. If any muriatic acid remains, it will continue to break down the mortar.

Brick floors should be sealed, but not until all the moisture has left the bricks and the mortar joints. Wait at least a month. If you get impatient and seal a brick floor with moisture in it, the sealer won't bond. The best product I've used for sealing brick is Hydrozo Water Repellent #7 (Hydrozo Coatings Co, Box 80879, Lincoln, Neb. 68501). Five gallons will do the average floor and will cost a little over $80. Put on two coats, letting each one dry for 24 hours. I follow this with a coating made of equal parts of turpentine and Valspar polyurethane varnish (The Valspar Corp., 1101 South Third, Minneapolis, Minn. 55440). □

Quarter-step
running bond

Third-step
running bond

Brick Floors

How a New Mexico pro lays a brick floor
without mortar or string lines

by Douglas Ring

Brick floors are common in adobe houses here in New Mexico. The deep-red earth tones typical of bricks and their slightly erratic dimensions fit right in with adobe houses. But these aren't the only structures in which I lay brick floors. Some clients want them because they make good heat sinks in passive-solar homes or good radiators in buildings with buried hot-water radiant-heating systems. Still other clients want them because of their looks. After all, one of the most popular sheet-vinyl patterns imitates the look of a brick floor, and the installed cost for the two materials is roughly the same—about $2 a square foot. I do point out to my clients that the two materials wear at different rates. A vinyl floor will probably need replacing in 20 years. But after 20 years the corners which stood a bit high in a new brick floor will be nicely rounded, and if it's been properly finished it will have a patina like a well-used banister. In heavy traffic areas maybe ¹⁄₆₄ in. will be worn away, leaving another 14 centuries of wear in those places—give or take a century or two.

Choosing the bricks—The phone book will tell you where bricks can be had in your area, and I strongly suggest that you visit each yard to assess the quality and price of the available bricks. Check the bricks for chipped corners and variations in dimensions. Up to 5% chipped corners is typical. If their dimensions fluctuate by more than ³⁄₁₆ in., you'll be spending an inordinate amount of time fitting them together.

We are fortunate to have a local factory (Kinney Brick Co., 100 Prosperity Ave. S.E., Albuquerque, N. Mex. 87102) that makes an inexpensive ($.17 each) low-fire brick suitable for interior use. Most of the brick manufacturers make their pavers (bricks without holes) for exterior use, which means the bricks have to be fired at higher temperatures. This makes them resistant to water so they won't spall when they are exposed to a rain followed by a freeze. Not surprisingly, high-fire bricks are more expensive than their low-fire counterparts because it takes

Douglas Ring is a licensed contractor in Albuquerque, N. Mex.

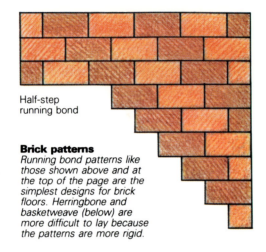

Half-step
running bond

Brick patterns
Running bond patterns like those shown above and at the top of the page are the simplest designs for brick floors. Herringbone and basketweave (below) are more difficult to lay because the patterns are more rigid.

Herringbone

Basketweave

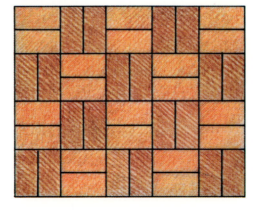

more time and gas to cook them. Around here, you can count on spending $.30 to $.50 apiece for high-fire bricks.

The brick you choose should have a smooth surface (roughness is a maintenance problem) and a pleasing color. Dark bricks will absorb more energy from the sun when they are used for solar gain. My favorites are light red or orange, and I keep in mind that the color will darken a little when the bricks are sealed.

Your supplier can tell you how many bricks you will need if you can tell him how many square feet you want to cover. If you want to calculate this number yourself, figure that to cover 1 sq. ft. it takes 4.5 bricks that are 4 in. by 8 in., or 5.2 of the 3⅝-in. by 7⅝-in. bricks. I add 5% to the total to allow for waste, and slightly more for a herringbone pattern. These figures are for bricks that are laid tightly together, without mortar joints between them.

I don't use mortar in my floors for several reasons. Mortar is messy and time-consuming to install, and it will permanently stain porous bricks. It doesn't strengthen the floor, and the mortar joints are slightly lower than the bricks, which makes grooves that collect dirt.

Every brickyard I've ever dealt with delivers their products. The trick is to have them deposit the bricks as close to the job as possible. The bulk of the work involved in laying a brick floor is moving the bricks.

A sand setting bed—Most of the brick floors I lay are atop concrete slabs. On top of the slab I screed a layer of sand, which can be made more level than the slab usually is, as well as accommodate the slight irregularities in brick thickness. The bricks I use are 2¼ in. thick, and I allow about ½ in. of sand between the bricks and the slab as a setting bed. This makes the distance from finished floor level to the slab 2¾ in. Of course, the slab can be more than ½ in. below the bricks, but that means you'll have to move that much more sand around to level your setting bed.

If I'm setting a floor over raw earth, I either remove or compact any soil that has been disturbed, and I fill in any craters with sand. Conversely, if grade is too high I'll take out soil until I'm about 3 in. below finished floor height. Here in New Mexico, moisture coming through the floor isn't a big problem, so we don't have to worry about elaborate drainage systems and moisture barriers under the floor unless the house is cut into a hill. Elsewhere, I expect these precautions would be critical.

Before I start bringing in the sand, I determine the finish-floor height from adjacent floors, door-sills or other fixed landmarks. Where possible, I snap chalklines on walls around the perimeter of the room to indicate the level of the finished floor. Then I add enough sand to bring it to the appropriate level below this line. The line is a big help in making sure that the floor is nice and level where it meets the baseboard. For the middle of the room, I rely on a hand level.

Sand is cheap, but it is also heavy. If I need a lot of it, I have it delivered along with the bricks. If I need more sand I know that I can carry enough to cover 200 sq. ft., 1 in. deep, in my pickup. That equals about ¾ ton.

Screeds and tongs—The best time to lay a brick floor is after the walls of the house have been painted—it isn't easy to remove paint from bricks. Once I've got the sand distributed around the room, I screed it level with the help of metal shelf brackets. These brackets are the kind that are U-shaped in section, with slots to accept the shelf standards. I lay a pair of the brackets in the sand and adjust them to the desired elevation with a level. Then I drag a straight board across two of these level and parallel brackets (photo top). I usually screed about a 4-ft. to 5-ft. wide path across the room in the direction the rows of bricks will run. Screeding done, I remove the brackets from the sand. They occupy so little space that I don't have to add any sand to fill in the grooves. I just smooth them over.

For every 100 sq. ft. of floor you lay, you will have to move about a ton of bricks. The tool you should have for this is a pair of brick tongs. These are like ice tongs only instead of a sharpened point to pierce the ice, they have a metal plate to grip a short row of bricks (photo top). The time they will save you moving only 500 bricks (100 sq. ft) will pay for them. If you have a great distance to move the bricks, a flat wheelbarrow is useful. You can make one of these by replacing the bucket of a regular wheelbarrow with two pieces of plywood. The flat wheelbarrow makes it easier to use the brick tongs to load and unload it.

To shape the odd-size bricks at the end of the runs you will need a 4-in. brick chisel and a 4-lb. sledge with a short handle. To position the bricks you will need a rubber mallet.

Running bond pattern—For beginners, I recommend some variation of the running-bond pattern (drawings, top of facing page). But just because this is an easy pattern to lay down doesn't mean that it isn't attractive. I've been laying brick floors full time for six years now, and this is still the pattern I use most. My favorite is the running bond based on fourths. I call it quarter-step running bond. This means that the ends of the bricks are staggered by a distance

Ring uses metal shelf brackets as screeds to level the sand bed (left). He adjusts each one to the desired height, then uses them to support a straightedge dragged across their tops to level the sand. The brackets are then removed, and the sand is smoothed over. Note the brick tongs being used by a helper. They are well worth the expense if you've got a lot of bricks to carry about.

The herringbone floor shown below intersects the wall and the neighboring course of bricks at a 45° angle, which makes it a floor with a lot of meticulously cut bricks. To make the bevels for the starter course, Ring lops the corners off a batch of bricks and then uses the corners to fill in the triangular voids.

equal to one-quarter of their length. The effect of this arrangement is subtle. It creates a repeating zigzag pattern that I think is more complex and interesting than the standard running-bond pattern based on halves. It's also possible to introduce another geometric pattern into a field of running bond, as shown in the bottom photo.

Most brick floors are oriented so the long dimensions of the bricks are perpendicular to the longest walls in the room. This is to avoid the frustration of trying to make long rows of bricks parallel to the walls. Variations in the bricks make perfectly parallel rows hard to achieve.

Assuming you use the quarter-step pattern, you will need equal numbers of bricks cut into quarters and halves and three-quarters for starter bricks. One brick can provide quarter-length and three-quarter-length starters for a row. The cuts don't have to be perfect because a baseboard will cover up any cuts right next to the wall. If you don't have a baseboard, make sure that the top edge of the brick is a crisp cut (see the sidebar on the facing page). It doesn't matter what the cut looks like below the surface.

Begin laying a brick floor by placing the starter courses in the corner of a room on a screeded bed of sand, as shown below. These starter pieces are the beginning of a quarter-step running-bond floor. The bottom photo shows how you can highlight a geometric pattern in a field of bricks by using bricks of different colors.

Setting the bricks—Cut a variety of starter pieces and begin to set them on a screeded setting bed in the corner of a room (photo below left). Don't worry too much about getting the size of the pieces exact. This is a handmade floor, and I think it's a waste of time to try and make it look like printed linoleum. First put a whole brick in the corner and tap it gently with the rubber mallet straight down—about 1/8 in. to 1/4 in. Then put a 3/4-length brick next to it, followed by a 1/2-length brick and then a 1/4-length brick. Then the sequence starts again. Do not tap the small pieces hard or they will sink too far. As you add more bricks to these rows, slide them straight down against their neighbors so that sand is not trapped between the bricks.

Keep adding rows of bricks, several rows at a time, screeding more sand when you run out of a flat place to lay them. When you finish laying all the bricks you can fit into the room, you will have two adjacent wall edges finished, and two walls with cuts still to be made. Do the cuts now. Waiting to do these cuts at the end of the entire job is like waiting until the end of dinner to eat your lima beans.

Aesthetics and adjustments—Before you tap each brick into place, make sure it doesn't have any unsightly nicks in it. If it does, turn it over or put it aside for cutting. You will have to use your judgment about what sort of defect is acceptable, as almost all bricks are a little flawed. I often save the worst ones for the parts of the kitchen that I know will be covered by cabinets.

As you lay the bricks, try to have a sense of what is level. Think of something like a calm ocean. If you try to level each brick with an instrument, the job will be tedious. If the sand is level, approximately the same amount of tapping will be sufficient for each brick. If you have to pound to get the brick even with its neighbor, chances are the floor is heading downhill. If this is happening, tuck some sand under the last brick with your fingers, fill any hole with a small handful of sand and continue. Have a level on hand but don't be a slave to it. Adobe building tradition in the Southwest encourages a handmade look, and a brick floor fits right into this sensibility. A more formal house wants a flatter floor, especially in the dining room. But how a brick fits with its neighbor is more important than whether or not one end of the room is 1/4 in. higher than the other.

Not only can you tip a brick accidently, you can also tip them on purpose. You can use this to advantage to change levels from one room to another to compensate for misplaced door sills, or even to make ramps instead of steps.

The straightness of the rows of bricks is another matter for individual interpretation. Most bricklayers want to get out the string and follow the straight line to ensure parallel lines. This is okay, and you will certainly achieve straighter rows of bricks by using a string line. But this is not required, once you realize that straight does not necessarily mean better. Most brick floors curve a little because of variations in the bricks and the walls. It adds to the charm.

Often you can curve the lines of bricks to good advantage. I've done floors in rectangular

rooms where I started the bricks in one corner and went to the opposite corner with a gentle S-curve. Then I filled in the rest of the floor, maintaining the gentle curves. On a floor like this, keep the curves smooth and large. Tight curves make bigger spaces between the bricks.

There is no rule that you must use the same pattern throughout a single house. In the bottom photo on the previous page, you can see where the pattern changes from quarter-step running bond to herringbone. Basketweave and herringbone are traditional patterns for brick paving that are more difficult than the running bond because the bricks are locked together in two directions. While both basketweave and herringbone patterns arrange bricks perpendicular to one another, herringbone is particularly challenging because of all the 45° cuts involved at walls or other courses of bricks.

Finishing and maintenance—When the last bricks are in place, straighten any rows that you find offensively crooked by twisting a trowel in the cracks, or by replacing oversized bricks. This is also the time to lift any bricks that are too low. Using two trowels to pinch a brick from the sides, lift it out, and add a bit of sand to the bed to make it flush with its neighbors.

Once you've made the necessary adjustments, sweep fine sand into the cracks between the bricks—an average-sized room will take about three or four shovelfuls. This is an exciting part of the job. The sand filters into the cracks as though the floor were a giant hourglass, locking the bricks tightly in place.

The floor is now basically complete. Interior floors should be sealed to resist staining. Standard practice has been to coat brick floors with liquid plastics, but my experience with refinishing floors has brought me to the conclusion that it is best to seal the bricks with something that penetrates deeper than plastic. This way you walk on the brick itself, which is almost indestructible, instead of on a thin layer of plastic. I mix my own sealer for this purpose (Ring Brick Floor Sealer, available from Ring Brick Floors, 2631 Los Padillas S.W., Albuquerque, N. Mex. 87105). One gallon costs $15, plus $5 handling, plus UPS shipping, and complete instructions for use are on the can. The mixture consists of about 80% linseed oil, along with thinners to help it penetrate and some additives to help it dry. This concoction penetrates the bricks to a depth of about 1/8 in.

For normal maintenance, wash the floor with a mixture of water and vinegar—about 1/4 cup vinegar per gallon of water. Don't use soap because it can leave a residue. Give it regular sweepings with a large dust mop sprayed with a conditioner like Velva-Sheen or Conquer-Dust (available at janitorial-supply stores). The dust-mop conditioner will keep the floor from looking dull but will not build up like wax. Liquid wax is a curse upon brick floors. It builds up and gets milky and dark. For a slightly higher shine, use Indian Sand Treewax (available at hardware stores). It is a brick-colored paste wax that will not turn milky or yellow. It won't build up because it is too hard to apply. Dust-mop maintenance is the same regardless of finish. □

The bevel on the tip of the brick chisel should be on the waste side of the cut, and the shank of the chisel should be tilted slightly away from the workpiece. In this position, it will create a clean edge and an undercut in the finished piece. Here Ring uses the removed portion, inverted and trimmed, to complete the angled face for a starter course in a herringbone floor.

Cutting bricks with a chisel

The fastest and most practical way to cut a large number of bricks is with a 4-lb. hammer and brick chisel. For what it would cost you to rent a diamond saw for one day you can waste about 100 bricks, and by the time you have cut 100 bricks you will be pretty good at making pieces. I make the larger pieces first so if I break the piece that I am making, I can make smaller pieces from the fragments.

If you are new to cutting bricks, start by making half-bricks. Put a brick on a small pile of sand so that it has a solid support. This will keep the brick from breaking from the bottom up because of a lumpy seat.

You don't have to measure for the cutline. Just place the chisel in the middle of the brick, perpendicular to its length, and whack it hard with the hammer. Watch what you are doing—mistakes can be painful. When you have mastered this (halves are easy), try one-third/two-thirds and one-quarter/three-quarters. The brick will break slightly away from the bevel of the chisel, making the piece on the flat side of the chisel more suitable for use in a floor. This undercut piece is less likely to have outcroppings protruding beyond the plane of the top edge. Since only the top edge will show, it doesn't matter what the rest of

the brick looks like as long as it doesn't stick out.

Bricks like to break across smaller cross sections and nearer to their middles, so rather than make a 45° cut all the way across a brick to start a herringbone course, I just take off a corner, as shown in the photos above. I use the triangular piece that I have

removed from the brick to complete my 45° bevel.

If the triangle comes out with excess brick on the underside, I trim it by chiseling it from the bottom, and I direct my blows so the pieces break off short of the top.

When you get to the end of a run of herringbone, you will often

To cut a brick with a long beveled edge, begin by shortening the brick to about 1/8 in. longer than the finished length. The second and third cuts remove the bulk of the remaining waste. The fourth cut makes a finished edge along the desired line, but because the cut is longer than the blade of the chisel, a fragment usually protrudes near the point. The fifth cut removes this protrusion, resulting in the beveled brick shown above.

need a long beveled piece to fill in the remaining gap (photo below). After much trial and error I have arrived at a sequence of cuts that will usually get me this piece.

First I cut the brick to near its finished length, which reduces its mass at the crucial area of the cut—the tip. I leave it about 1/8 in. long, because it often chips a bit at the corner. The second cut should go right to the tip and can be slanted back to start paring down the remainder. Too much angle here will lose you your point. The third cut trims away more brick. The fourth cut is on the finished line, with the flat of the chisel always toward the piece you want. The final cut is one sharp blow to cleave away the remaining fragment near the point.

For cuts the full length of a brick, don't hit the chisel hard enough to break the brick through the first time. Whack it on both ends with increasing firmness until it breaks. You can even hit it on the end to start it cracking the way you want. If ten blows have not opened it up, hit it harder. If the piece you want is the length of the brick and less than 1½ in. wide, it is better to use two short pieces. An extra joint in the floor or even an extra row of joints is "authentic."

—D. R.

Brick-Mosaic Patios

Creating patterns and pictures with dry-laid brick and a diamond-bladed saw

by Scott Ernst

"That was a memorable day to me, for it made great changes in me. But it is the same with any life. Imagine one selected day struck out of it, and think how different its course would have been. Pause you who read this, and think for a moment of the long chain of iron or gold, of thorns or flowers, that would never have bound you, but for the formation of the first link on one memorable day."—Charles Dickens, *Great Expectations*.

That day came the summer after I graduated from college with a degree in landscape architecture. I had been doing dry-laid brickwork for nine years, frequently working curves into my patio designs. And when you make curves with brick, it usually means a bit of cutting. So I was doing just that, one warm June day, when my design background suddenly collided with my years of brickwork, and a strange idea popped into my head: if I took the piece of brick that I had just cut off and replaced it with an identically shaped piece from a different colored brick, I could create patterns and pictures.

A few months later I produced my first piece of "masonry art"—a form that's similar in concept to stained-glass, mosaic tile and marquetry. That was five years ago, and since then I've come to specialize in brick mosaics (photos right). While the limitations are considerable, there's still room for a lot of creative freedom...and a lot of cutting.

Brick colors and types—After stumbling upon this concept, the first thing I had to do was find out what palette of colors was available. I spent many an hour strolling through brickyards, making mental notes on who had what colors and in which sizes. To date, my list includes basic red, several tones of brown and tan, grey, blue, rust and white. Many of these bricks also come "flashed," which means that they have a range of colors or tones within each brick variety.

The trick is to find the colors I want, all in one size. An assortment of brick sizes won't make a tight pattern and cutting them to a uni-

This patio, depicting a heron scene that Ernst calls "Rising," is composed of variously colored brick, cut and pieced together on a dry base. Ernst admits that the technique takes a lot of time.

form dimension is an excruciating job. Because of the need for consistent sizes, I like to get all of my bricks from one manufacturer. I use Glen-Gery bricks almost exclusively (Glen-Gery Corp. 1166 Spring St., P. O. Box 7001, Wyomissing, Pa. 19610; 215-374-4011). Their bricks are high quality and come in a wide variety of colors.

As a rule, the lighter the color of the brick, the denser it is, making it harder to cut. The used to make these bricks are finer and clays

contain less grog than the darker colors. These different properties allow me to make certain cuts with some bricks that just aren't possible with others. Harder bricks, for example, hold together better for fine detail work such as the grasses in the heron scene (photo below).

Bricks come in two basic types: molded and extruded. Each has its own look and application. Traditionally, bricks were shaped by slapping a bunch of clay into a mold, striking off the excess and letting it harden. This produces a fairly irregular brick, which when laid into a patio, tends to leave large joints between the bricks. This looks nice if you want the patio to appear old. For most inlays, however, molded bricks are not the brick of choice.

To keep the lines in a design crisp, I want bricks to fit together as tightly as possible, so I use extruded bricks for my inlays. These bricks are wire-cut instead of molded and thus are more uniform in shape and size. They are also more difficult to cut than molded brick. (For more on brick manufacturing see *FHB #53* pp. 51-53.)

A third type of brick is the core brick, which has holes in it and is designed for building walls. Core bricks can be used as a textural element in mosaic designs, but I've only used them indoors. I have some reservations about how well they would stand up to freeze/thaw cycles outdoors.

Including a bit extra for breakage and cutting, there are about 4.7 bricks per square foot. I multiply this by the number of square feet of each color in the design to come up with the number of bricks I need.

Design considerations—The critical limitation here is that you cannot blend colors; everything must be drawn in blocks of colors or tones. It helps to study high-contrast black and white photos of your subject matter when designing a piece. These can often be simplified to two or three defined tones and turned into cutouts.

To a limited extent, it's possible to create color tone gradations within a brick inlay if

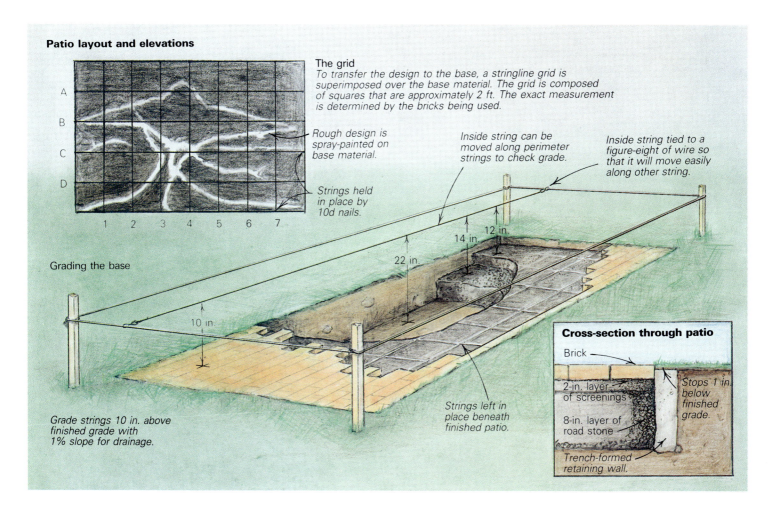

Patio layout and elevations

A
B
C
D

1 2 3 4 5 6 7

The grid
To transfer the design to the base, a stringline grid is superimposed over the base material. The grid is composed of squares that are approximately 2 ft. The exact measurement is determined by the bricks being used.

Rough design is spray-painted on base material.

Strings held in place by 10d nails.

Inside string can be moved along perimeter strings to check grade.

Inside string tied to a figure-eight of wire so that it will move easily along other string.

Grading the base

10 in.

22 in.

14 in.

12 in.

Strings left in place beneath finished patio.

Grade strings 10 in. above finished grade with 1% slope for drainage.

Cross-section through patio

Brick

2-in. layer of screenings

8-in. layer of road stone

Trench-formed retaining wall.

Stops 1 in. below finished grade.

you use a flashed paver. The various tones in a pallet of flashed brick can be separated into piles of light and dark tones. Then when they are placed into the pattern, the bricks are used selectively to produce the effect of a gradual change in lighting or simply to highlight another block of color that overlaps the first. This technique does not work well using two different varieties of brick because they are not often in the same color family and will look out of place when blended together.

Also in the back of my mind when designing brick inlay is the fact that the bricks are going to get dirty and the contrast between colors will be reduced over the years. So when I choose the brick colors, I'm a little heavy-handed with the contrast; things will tone down later.

Another approach to the dirty-brick problem is to waterproof the bricks. The theory is that if no water gets into the pores, no dirt will get into the brick. This is something that I recently tried on a civic project in Reading, Pennsylvania, where I did a map of the city in brick. But it will be several years before I know how well the waterproofing works.

Base preparation—A variety of base materials can be used for dry-laid brick. Among them are sand, stone-dust screenings (also called quarry process) and fine gravel. I long ago quit using sand under my patios. It just isn't stable enough for my taste. It isn't bad; there are just better options.

I use screenings most of the time. This is the stuff left over when stone is crushed and gravel is screened out. The size of the chunks ranges from ⅜ in. down to a fine powder. The large pieces give stability, and the powder fills in the gaps. When it is wetted and tamped, the base becomes very stable and can be walked on without disturbing the grade.

Another good option for base material is ¼-in. gravel. It can be compacted quickly and provides extremely good drainage, which keeps root intrusion from nearby trees to a minimum. The problem is that it's hard to find (at least here in central Jersey).

For an average dry-laid patio, a 4-in. base will last 15 to 20 years before it needs a bit of regrading. The Romans used a 6-ft. base under roads that are still in place. I kind of hedge my bets. If I'm going to sink four or five months of my life into a patio, I'd like to see it last longer than 20 years, so I use 10 in. under my inlaid patios. Rather than use a full 10 in. of screenings, however, I put down 8 in. of road stone (screenings with ¾-in. to 1-in. gravel mixed in) and cap it with screenings. The road stone gives the base an added measure of stability. A deep base must also be tamped in lifts (layers) of about 3 in., or you won't get good compacting.

Another method of base prep is to pour a 4-in. to 5-in. reinforced-concrete slab, leaving room for ¾ in. to 1 in. of screenings on top. This method is more common for commercial installations that get lots of traffic.

Methods of base prep not applicable to this type of inlay are wet laying and drypack (lay-ing the brick on a dry mortar base and wetting). The reason for this, as you will see later, is that the bricks are all placed and then individually lifted and cut. If these methods were used, the bricks would be stuck to the ground and I wouldn't be able to cut them.

The actual grading of the base is no different than standard dry-laid work. For most residential patios, I use four stakes, connected by stringlines set 10 in. above finished grade and set at about a 1% slope for drainage (drawing above). Ten inches puts the strings at a convenient height so I can work over or under them. I run a fifth string between the two longest sides, tied so that I can slide it along them to check the grade anywere inside the perimeter.

I dig 22 in. below the stringlines, then start adding base material, which I level roughly with a rake and compact with a 10x10 hand tamper (I'll rent a power tamper for big jobs). I tamp 3-in. layers until the base is about 10 in. deep; I do the final grading with a 2-ft. and then a 6-ft. board, screeding the base like concrete.

From paper to brick—After I superimpose a scaled grid onto both my final drawing of the design and the base material, I use the grid to enlarge the drawing. For projects that are highly detailed, a 1-in. grid on the drawing and a 2-ft. grid on the base seem to work well. A larger grid will be fine for small or less-detailed designs.

I lay out the grid on the base with strings tied to 10d nails pushed into the base material (drawing above). Both the grid on the drawing

Extra effort. **After tracing the S-shaped pattern onto the brick (1), Ernst makes a series of cuts on a tile saw (2) and knocks out the pieces with a hammer and chisel (3). Then he smoothes the edges on the tile saw and traces the negative shape onto the next brick (4).**

and the grid on the base are then labeled A, B, C, D, up one axis and 1, 2, 3, 4, across the other, giving me reference points to work from while transferring the design.

It is important to note here that different varieties of brick that are theoretically the same size usually are not. For example a Glen-Gery 4x8x2¼ "Chambersberg" paver is always a hair smaller than Glen-Gery's "Longwood," which is truly 4x8x2¼. If I were laying either of them by themselves, this size difference wouldn't be a problem, but when placed tightly together into a patio, the pattern begins to wander. For this reason my string grid is doubly important. I lay it out in modules based on the largest brick.

I determine these modules by laying out a test pattern of the largest brick and measuring it. In bricklaying, unlike mathematics, the sum of the parts does not equal the whole. This is because no matter how tightly you lay your pattern, there is always a bit of space between the bricks. Consequently the actual grid will be anywhere from 23½ in. to 24½. The rest of the brick varieties will then be spaced to fill the grid. I don't fret about these gaps, though; they're usually minimal.

Using the numbers and letters on the grids

as reference points, I now go grid by grid and spray paint the design onto the base, using the cheapest paint I can get. The screenings are dark, so I use white paint. Detail is not important here—that will be worked out later. This step is meant only to give me proportions for laying the bricks.

Next comes the easy part. I start in a corner with the color that belongs in that corner and begin laying my pattern. When I hit one of the spray-painted lines, I begin laying the color that belongs in that section, then the next, etc., until the whole patio is laid. At this point the design will look somewhat like an enlarged computer image, only with bricks instead of pixels.

Now I draw the finished design onto the bricks, using the blocks of color as a reference for the proper proportions. Fine details that were hinted at in the drawings now must come to life. Compositions within the design when they were 1-in. tall may need to be redrawn when they swell to two feet.

Chalk works well for the initial sketching as it erases easily with a large, damp sponge. Once everything is finalized, though, the drawing gets hard-lined with a more permanent marking (photo above left). China markers,

grease pencils and some brands of colored pencils all work for this. I prefer Prismacolor pencils because they go on and stay on well, have vibrant colors that contrast with the colors of the brick and give a nice, fine line.

Tool tech—The machine I've found that cuts curves in bricks most efficiently is a 10-in., water-cooled circular saw with a diamond blade. I've experimented with a few brands and my favorite is the Target Tile Saw (Target Products, Inc., 4320 Clary Blvd., Kansas City, Mo. 64130; 816-923-5040).

Blades come in two basic styles: segmented and continuous rim. The continuous rim is just that; it looks like a blade with no teeth. The outside edge of the blade, though, is embedded with diamond bits. A segmented blade works on the same idea, but it has cutouts that break the rim into segments. This is the superior blade type if you're going to do a lot of cutting. The segments add a bit of extra bite that increases cutting speed and blade life.

Diamond blades are made for a number of applications, one of which is brick-cutting. But of course, it is isn't *that* simple. Blades are available for hard brick and soft brick,

Beautiful results. Although brick mosaics entail a lot of cutting, the results, like the tree shown on the facing page, are certainly beautiful. Experimenting with a diamond-bladed bandsaw, Ernst cut out the jolly message at left to photograph for his Christmas cards.

long life and fast cutting. I've been pleased with NED-KUT (N-E-D Corp., 18 Grafton St., Worcester, Mass. 01604; 508-798-8546) and Pearl blades (Pearl Abrasive Co., 6210 Garfield Ave., Commerce, Calif. 90040; 213-773-1625), but have had problems with Target blades.

You might be wondering why this knucklehead doesn't just go out and pick up, oh, a diamond-bladed bandsaw, or something like that. After all, a bandsaw cuts curves better than a circular saw. Well, I thought the same thing. So I picked one up and discovered, much to my dismay, that the blade just didn't cut brick very fast. Still, the bandsaw is great for certain highly detailed projects, like the "Season's Greetings" mosaic that I did as a Christmas card (photo above).

Cutting, and more cutting—First of all, I always wear a full face shield, hearing protection and a respirator when I'm cutting brick. Once I've suited up and am ready to cut, I take a stout screwdriver and pry out one of the bricks slated to be cut. If the cuts are straight, they're easy to make. Put the brick on the saw, roll it under the blade, and wham, it's cut sweet as can be. If I'm cutting curves, though,

the process is a bit more complex. External curves with the circular saw aren't too bad; I simply make a series of straight cuts tangent to the curve (top right photo, facing page).

Internal curves are more time consuming. For these I make parallel cuts into the curve anywhere from from ⅛ in. to ½ in. apart (as though cutting teeth for a comb). This is complicated by the fact that the blade is above the work surface, like a radial-arm saw, instead of coming out of the work surface like a table saw. This means that none of the cuts are square; they're all over-cut. To correct this I lift the leading edge of the brick and point it at the center of the blade while cutting so that the top and bottom are cut equally. While this is a commonly used technique among brick masons and tile setters, it should only be performed by an experienced tradesman, and some portion of the brick should *always* be in contact with the table. After making the cuts, I knock out the teeth with a cold chisel or hammer (photo, top left) and grind the cut smooth with the edge of the saw blade, again tilting the brick up.

The basic idea is that two colors are meeting along a line that I've drawn. The brick in

my hand is one of those colors, and I want to cut off the part of that brick that is in the way of the other color. Later I cut a piece of the other colored brick using the first piece as a template (photo above right). When I fit it against the first piece, I'm back to the original brick dimensions. I drop this unit back into the patio and pry up the next brick to be cut. After doing this for a few months, I've got an inlaid patio (bottom right photo, facing page).

Once the brick is all laid, I dig a trench around the perimeter of the patio, usually 10 or 11 in. deep and 4 in. wide, which I fill with concrete mixed on site (drawing, p. 87). The dirt serves nicely as formwork. I trowel off the concrete about 1 in. below finished grade; this way it serves to retain the brick but can still be covered with topsoil and grass. As a final step, I buy bags of play sand (which is drier than sand I can get from a masonry supplier), broadcast it on top of the patio and sweep it into the cracks to lock the bricks into place. □

Scott Ernst is an itinerant craftsman who lives in his truck and specializes in brick mosaics. He can be contacted at P. O. Box 831, Dayton, N. J. 08810.

Laying Up Brick Bovedas
Inwardly leaning arches defy gravity

by E. Logan Wagner

It's not often that an architect gets a chance to incorporate traditional masonry domes, called *bovedas,* into a house, but that opportunity came to me. My clients owned a site on the north slope of a 5,000-ft. mountain in hot, semi-arid southwestern Texas, 80 miles north of Big Bend National Park. They wanted a house incorporating traditional elements of Spanish architecture, so the final design centered around an open courtyard. Square in plan, each corner of the house was topped by a boveda (photo right).

A boveda is a dome built without the aid of formwork. Made of brick or other masonry material, bovedas originated in Egypt and the Middle East, where ingenuity made up for a lack of trees to provide lumber for roof supports. During the Renaissance, masonry domes took on an unprecedented popularity and were standard fare for most religious buildings in Italy and Spain. And when the Spanish conquered the Americas, missionaries converted the natives first to Catholicism and then to Spanish masonry techniques.

Many churches were built with bovedas, as were granaries, kiosks and water cisterns. Far from dying out when the Spanish left, however, the rich tradition of masonry bovedas survives to this day, primarily via a handful of skilled masons from the eastern part of the state of Jalisco, Mexico. The house built for my clients called for considerable skill on the part of the builders, but the star of the show was the *bovedero,* Sr. Don Alfredo Avila Almaguer, son of Don Mateo Avila (see "O'Neil Ford's Boveda House," *FHB* #23, pp. 26-31).

A bovedero is a mason who specializes in the making of bovedas. The traditional techniques of making bovedas were rediscovered in Lagos de Moreno, Jalisco, Mexico, by Don Alfredo's grandfather many years ago. As Don Alfredo recounts the story, his grandfather and a few other masons in Lagos were hired to remodel an old residence. While doing the work, they stumbled upon an abandoned cistern built in the boveda tradition. Intrigued, Don Alfredo's grandfather and his companions set out to try and recreate the boveda process. Two generations later, the building of bovedas has been returned to the level of virtuosity prevalent during colonial Mexico, and Lagos de Moreno has also become the boveda capital of the country.

The basic brick—The materials and tools needed for boveda-making are few. The brick, known in Jalisco as *ladrillo de cuna* or wedge brick, is a lightweight and relatively soft brick made from the clay soils around Lagos de Moreno. Mixed with water, the soil turns to mud and is poured into wooden molds, each containing four bricks. When released from the molds, the partially dry bricks are left to dry completely in the sun. Two good sunny days are sufficient for drying the bricks to a whitish appearance, whereupon they are fired in a wood-fueled oven for one day and one night.

The final size of each brick is roughly 2 in. by 4 in. by 8 in., and each one is slightly cupped due to the drying process. This cupping creates suction in the mortar that helps the brick, as it is being layed, to adhere to the course below.

The basic mortar—For mortar, Don Alfredo makes up a 4:1:¾ mix of sand, lime and portland cement, and adds just enough water to give it a thick, but manageable consistency. It hardens in a relatively short period of time, and combined with the suction created by the curvature of the brick, the mortar is able to hold bricks in place at steep angles without support. The porosity of the brick encourages an even faster bond by absorbing the moisture of the mortar.

Mixing the mortar is labor-intensive because Don Alfredo prefers not to use a mechanical mixer. First, he creates a volcano-like cone of sand, then adds the lime and finally the cement. A crater is formed in the "volcano" and then filled with water. Using a hoe and shovel, Don Alfredo skillfully cuts in

portions of the dry material so that the water does not spill out over the edge of the crater. As the mixing continues, more water is added as needed. Cement dye is sometimes mixed with the dry ingredients in order to lend color to the mortar.

The basic tools—The steel trowel is one of the bovedero's most important tools. With it, he is able to transport the mortar skillfully from a shallow wooden tray to wherever the next brick is to be layed. While this technique is perhaps familiar to masons the world over, a bovedero will also follow the unconventional technique of using his trowel to cut and shape the brick.

Bricks with one rounded end are used to form a cornice around the interior base of the boveda (see drawing below). Called the *pecho de paloma*, or pigeon's breast, the cornice thrusts 2 in. to 3 in. into the center of the room as decorative trim. To shape each brick, Don Alfredo holds it in one hand while chipping away at the edge on one end with his trowel. He works to no layout lines, and yet shapes each brick to a uniform, graceful curve. Then, with a coarse stone, the brick edge is smoothed out into its final form. In the hands of an expert, the procedure takes somewhere between 10 to 15 minutes per brick. For one with little experience, however, it is advisable to sketch out the finished shape of the pecho de paloma on the slender side of the brick, so that the curvature of all bricks is nearly identical.

There aren't many other implements in the bovedero's tool repertoire. He needs a steel brush to clean off excess mortar after several courses of brick have been set. A small wood wedge, made from scrap wood found at the construction site, is used to rake the mortar at each course. The mortar box is made from wood, and is elevated to a comfortable height so that the bovedero exerts the least energy in his delivery of mortar from the box to the boveda.

By definition, a boveda is layed up without benefit of formwork. Don Alfredo does, however, use scaffolding in order to reach the interior of the boveda. After carefully leveling the scaffolding on adobe piers, he planks it at a level just shy of the pecho de paloma. From this level the boveda will take shape, and its height will be determined by Don Alfredo's reach.

The structural support—The bovedas on this house each span approximately 18 ft., although Don Alfredo has built bovedas with spans of up to 40 ft. In any case, the weight of the dome is considerable, and proper support for the boveda is crucial. Exterior walls for this house are load-bearing adobe. Asphalt emulsion was added to the dirt mixture to make the adobe weather-resistant (this is called "stabilized" adobe). To reinforce the tops of the walls, a crew from Rainbow Adobe, builders and adobe brick makers in Alpine, Texas, formed and poured a continuous, reinforced-concrete collar beam (see drawing below). On the inside of this collar beam, a lip provides a platform from which the boveda will spring. The boveda can cover a rectangular, square or circular space, and the collar beam must be shaped accordingly.

After the collar beam has cured, one course of brick can be layed flat on the inner lip; this is the pecho de paloma, which trims the base of the boveda. Although the cornice is not structurally necessary, it gives the interior of the boveda a pleasing, elegant base. And as a practical matter for contemporary homes, it provides a recessed space for indirect lighting. Later on we installed low-voltage string lighting (similar to Christmas-tree lights) along the lip.

Starting the boveda—A boveda begins simply, but this beginning is one of the most crucial steps in the process of construction. Once the desired shape of the boveda is determined, the bovedero adjusts the angle of repose of the brick to form the curvature desired. The rounder and taller the dome, the greater the angle of the brick from the horizontal. Bovedas are a series of arches that "lean" on the arch layed previously. At each corner, progressively larger arches are layed

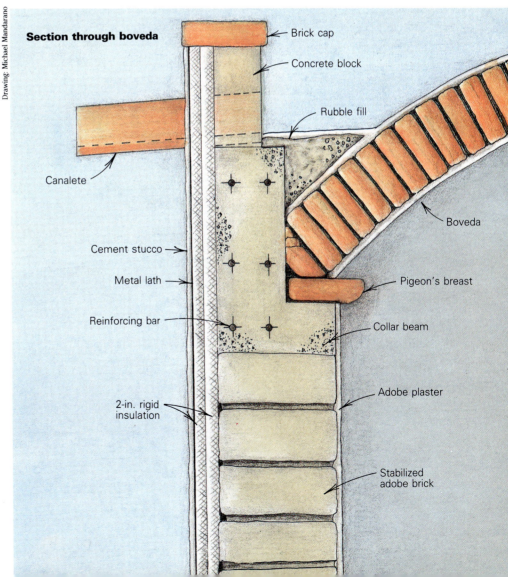

Drawing: Michael Mandarano

Section through boveda

- Brick cap
- Concrete block
- Rubble fill
- Boveda
- Canalete
- Cement stucco
- Metal lath
- Pigeon's breast
- Collar beam
- Reinforcing bar
- Adobe plaster
- 2-in. rigid insulation
- Stabilized adobe brick

The boveda begins on a ledge formed in a reinforced-concrete collar beam. The curve of the boveda shell starts rather simply with small fragments of brick embedded in thick mortar (photo above left). Additional, slightly larger, brick fragments are added to form an arch over the first few brick fragments (photo above right), and larger bricks are carved away where they meet the vertical face of the collar beam. Soon the bovedero is able to use full-sized bricks, and a segment of the boveda begins to take shape (photo left). Note the lack of formwork to support the dome. Due to suction created by a shallow depression in the underside of each brick and to the thick mortar, bricks will stay in place on the angled dome without any aid. The size of each segment of the boveda is determined in part by how far apart corners are. Adjoining segments meet halfway between two corners (photo below), and later on, the space between them will be filled with more bricks. At the apex of the dome, a small area is left open to provide space for the lantern. At this point, the dome is strong enough to stand on, and the bovedero seals the dome with layers of cement stucco and fiberglass-reinforced cement (photo facing page).

until adjoining arches meet halfway between their respective starting points.

Don Alfredo begins the boveda at one corner of the collar beam, and with his trowel, chops the brick into small pieces 1½ in. thick and the width of a brick. Three such pieces form a tripod that is embedded in a thick slop of mortar (top left photo, facing page). With slightly larger pieces, he forms a tiny arch supported by the tripod. The next arch is larger, but the span is not yet long enough to require the full length of a brick (top right photo, facing page). Each successive arch calls for longer and longer brick pieces, until full bricks are being layed (middle photo, facing page).

At the peak of each arch, a piece of brick must be fit exactly to fill the gap remaining between the two opposing rows and is, in effect, a keystone. So many odd and irregular brick pieces are used over the course of building a boveda that brick scraps are routinely collected for later use.

After each two or three courses, Don Alfredo stops to clean and screed the excess mortar from the coursing. While he has been working, his indispensable assistant has been very busy supplying him with a constant supply of mortar and bricks. The assistant also helps Don Alfredo to rake out the excess mortar between the brick courses, and cleans successive courses with a steel brush.

The partial vault at one corner is complete when the largest arch reaches halfway along adjacent sides of the collar beam. Don Alfredo then starts the process again in the next corner. Again he begins with small pieces of brick, the seedlings of the next vault, and patiently forms ever-longer arches. With two corners complete (bottom photo, facing page), Don Alfredo builds the remaining two corners, and finally, the four dome-segments stand ready for the next phase. Outlined against the sky, the corners

are strikingly beautiful, as if the rounded corners of the brick shell were flower petals yielding to the budding of a flower.

Completing the boveda—With all four segments of the dome complete, Don Alfredo begins to fill the spaces between them. He sets two courses of brick between each segment in the V-shaped intersection of two vaults, then repeats the process two courses at a time around the dome, gradually working upwards. Soon he has almost completed the full curvature of the boveda—only a small, squarish opening remains at the very apex of the dome, and it is from this point that the *linternilla* will rise (photo below).

The owners wanted to surmount the shell with a linternilla, or lantern. Windows in a lantern allow hot air rising from the rooms below to escape, and this encourages convective currents indoors. The lantern also plays a soft, natural light on the interior surface of the boveda and serves as an ornamental crown for the exterior.

By this time the dome, though unfinished, is strong enough to stand on, and Don Alfredo hikes to the top. Here he builds four brick stub walls to surround the small opening in the top of the boveda, and frames a window into each. Building the lantern walls requires modest skills, at least compared to the work preceding it.

Echoing the shape of the boveda, a domed cupola, made of adobe brick, is built atop the lantern. A cornice of 12-in. by 12-in. clay tile, called a *ceja* (eyebrow), provides the base from which to build the cupola. The first course of clay tile is guided by a string compass centered in a temporary wood frame spanning the interior of the lantern. Each circular course of the cupola becomes progressively smaller until the half-sphere is completed. To this point, Don Alfredo and his assistant have taken ten working days to build the boveda.

A hard shell and a parapet wall—Although the raw brick boveda has a most beautiful appearance, it requires a cohesive membrane that will consolidate the bricks in the shell and protect them from the weather. For these bovedas, two coats of cement stucco and a top coat of fiberglass-reinforced cement were applied to the larger dome of the boveda.

The cupolas called for a different approach. After three coats of cement stucco were applied to the cupola, it was painted with a coat of moisture-sealant paint. Then ceramic tile was laid in thinset exterior tile mortar (Custom Building Products, 6511 Salt Lake Ave., Bell, Calif. 90201). This mortar resists the large temperature changes typical of the high-country climate of southwest Texas.

A parapet wall 2-ft. high surrounds each boveda. Fill consisting of masonry rubble, covered by cement, creates slopes to conduct the water toward cut-stone water spouts called *canaletes*. The parapet wall keeps water from washing over the walls of the house.

The *boveda de aristas* —The basic construction techniques involved in making a brick boveda can also be put to other uses. A boveda that Don Alfredo is particularly proud of is his *boveda de aristas*, or groined vault. Located at the main entrance of the house, this boveda is supported by four brick arches, one for each side of its base. The columns (groins) of the vault form the arches and become wider as they rise, until eventually, the brick coursing fuses into the round boveda itself. This masterpiece is a fitting tribute to the exquisite virtuosity of brick and to the talented bovedero who builds with it. □

E. Logan Wagner, an architect and anthropologist, grew up in Mexico and now works out of Austin, Tex. Photos by the author.

Building a Fireplace

One mason's approach to framing, layout and bricklaying technique

by Bob Syvanen

I have been involved in building as a designer and carpenter for over 30 years, but building a fireplace has always been a mystery to me. I recently had the chance to clear up the mystery by observing, photographing and talking to my mason friend, John Hilley, as he built three fireplaces. I now understand more clearly than before what I should do as a carpenter and designer to prepare a job for the mason. I also know I can build a fireplace.

The job actually begins at ground level, with a footing (drawing, facing page). A block chimney base carries the hearth slab, upon which the firebox and its smoke chamber are built. The chimney goes up from there.

The importance of framing—As a carpenter, I've had to reframe for the mason too many times. This is usually because the architect or designer didn't realize how much space a fireplace and its chimney can take up, and how this can affect the framing around and above it. We'll be talking about a fireplace built against a wall, which is a pretty simple arrangement, but planning is still important.

Most parts of the country have building codes that specify certain framing details. In Massachusetts, where I live, code requires that all framing members around the fireplace and chimney be doubled, with 2 in. of airspace between the framing and the outside face of the masonry enclosing the flue.

The modified Rumford fireplaces that Hilley usually builds are my favorites because they don't smoke, they heat the room about as well as a fireplace can, and they look good. The firebox is 36 in. wide by 36 in. high, and the two front walls, or pilasters (returns) are 12 in. wide, for a total masonry width of 60 in. From the fourth course above the hearth, the rear wall of the firebox curves gently toward the throat. It's harder to lay up than a straight wall, but I think it looks a lot better. The back hearth is 20 in. deep and about 18 in. wide at the back—not in line with Count Rumford's proportions (see pp. 102-105), but the minimum allowed by the Massachusetts code.

To figure the full masonry depth, you have to add to the 20-in. back hearth 4 in. for the back-wall thickness, 4 in. for the concrete-block smoke-chamber bearing wall, and 4 in. for the concrete-block substructure wall, for a total of 32 in. Thirty-six inches is better, because it gives extra space for rubble fill between the back wall and the block. Using

these dimensions, the chimney base is 36 in. by 60 in. Add a front hearth depth of 24 in. (16 in. is minimum), and clearance of 2 in. on each side and rear, and you get a total floor opening that's 64 in. wide by 62 in. deep. In situations like this one, where the fireplace is on a flat wall and the chimney runs straight up, with no angles, the framing is simple—double the framing around the openings and leave 2 in. of clearance around the masonry.

To locate the flue opening in the floor above the fireplace, find the center of your layout and drop a plumb line. This determines the side-to-side placement of the flue. Its depth is determined by the depth of the firebox. The flue will sit directly over the smoke shelf, and is supported in part by the block and brick laid up behind the firebox's rear wall. The framing for the chimney depends on the flue size. An 8x12 flue requires a minimum 18x22 chimney (a 1-in. airspace all around, inside 4 in. of masonry). Once the ceiling opening is framed, you can establish the roof opening by dropping a plumb bob from the roof to the corners of the ceiling-joist opening.

Wood shrinkage is something you should take into account when you're framing around the hearth. I think the hearth looks and works best if it's flush with the finished floor. Since it is cantilevered out from the masonry core (see below), and isn't supported by the floor framing, shrinking joists and beams can leave it standing high and dry. I've seen fireplaces built in new houses where the 2x10 floor joists rested on 6x10 beams. The total shrinkage here could leave the hearth an inch above the finished floor. A better framing system is to hang the joists on the beams and thereby reduce the shrinkage 50%.

From footing to hearth—The fireplace really begins at the footing, which is usually a 12-in. thick concrete slab 12 in. larger all around than the chimney base, and resting on undisturbed soil. The footing for this fireplace, therefore, is 48 in. by 72 in. Between it and the concrete hearth slab is a base, usually of 8-in. concrete block if it is in the basement or crawl space. To make sure the hearth comes out at the level you want it, the height of this base has to be calculated to allow for the 4-in. thick reinforced-concrete hearth slab, the bed of mortar on top of it, and the finished hearth material—in this case, brick.

Before pouring the hearth slab, the opening

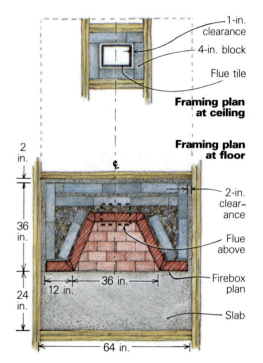

1-in. clearance

4-in. block

Flue tile

Framing plan at ceiling

Framing plan at floor

2-in. clearance

Flue above

Firebox plan

Slab

2 in.

36 in.

24 in.

36 in.

12 in.

64 in.

in the top of the concrete-block base is covered with a piece of ½-in. plywood that is supported by the inside edges of the blocks, leaving most of the course exposed for the slab to bear on. Cover the holes in the block with building paper or plastic, and build the formwork, secured to the floor joists, to support the cantilever at the front of the hearth. Then pour your 4-in. slab over a 12-in. grid of ⅜-in. rebar located 1 in. from the top.

Once the hearth slab has cured, it's time to lay up the structural masonry core that will support the chimney. Only the firebox, pilaster and lintel bricks will be visible on the finished chimney, so Hilley used 4-in. concrete block for the core. The blocks should be laid at least 4 in. from the face of the firebox brick and far enough in from the line of the front wall to allow for the pilaster bricks. Hilley sets a brick tie in each course to tie the pilasters in with the block.

Before beginning the brickwork, Hilley nails vertical guide boards (drawing, p. 96) to the face of the studs that frame the walls on each side of the fireplace opening, from floor to 12 in. above the lintel height. These boards are the thickness of the finished wall, and they locate the face of the fireplace. He marks off the brick courses up to three courses above the lintel on each guide board, starting from the

Illustrations: Christopher Clapp

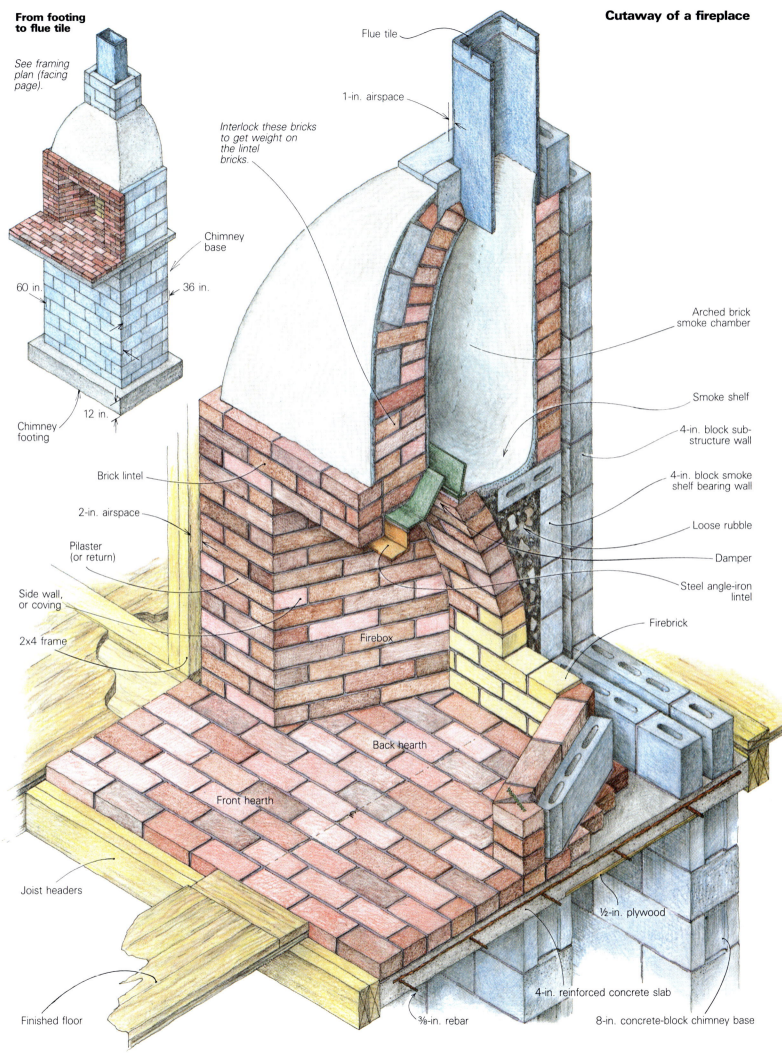

From footing to flue tile

See framing plan (facing page).

60 in.

36 in.

12 in.

Chimney base

Chimney footing

Cutaway of a fireplace

Flue tile

1-in. airspace

Interlock these bricks to get weight on the lintel bricks.

Brick lintel

2-in. airspace

Pilaster (or return)

Side wall, or coving

2x4 frame

Joist headers

Finished floor

Firebox

Back hearth

Front hearth

Arched brick smoke chamber

Smoke shelf

4-in. block sub-structure wall

4-in. block smoke shelf bearing wall

Loose rubble

Damper

Steel angle-iron lintel

Firebrick

½-in. plywood

4-in. reinforced concrete slab

⅜-in. rebar

8-in. concrete-block chimney base

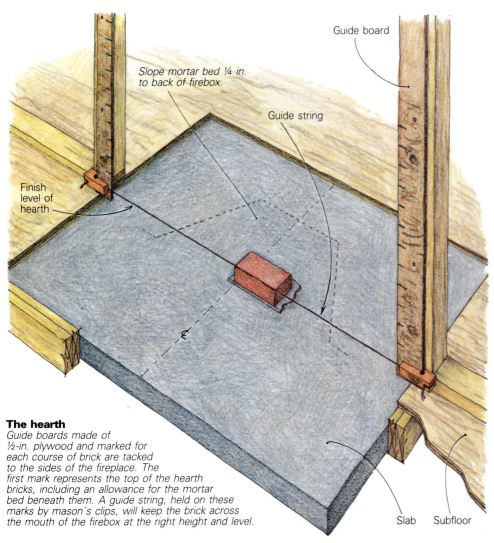

Slope mortar bed ¼ in. to back of firebox.

Guide board

Guide string

Finish level of hearth

The hearth
Guide boards made of
½-in. plywood and marked for
each course of brick are tacked
to the sides of the fireplace. The
first mark represents the top of the hearth
bricks, including an allowance for the mortar
bed beneath them. A guide string, held on these
marks by mason's clips, will keep the brick across
the mouth of the firebox at the right height and level.

Slab   Subfloor

Firebox layout

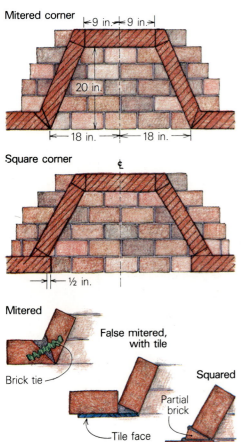

Mitered corner

9 in. | 9 in.

20 in.

18 in. | 18 in.

Square corner

¢

½ in.

Mitered

Brick tie

False mitered, with tile

Squared

Partial brick

Tile face

Corner details

The firebox is being laid to the penciled layout, starting with five courses of the back wall. Notice the curve starting at the fifth course of the back wall. The cut-brick piece for the front mitered corner will be alternated from front wall to sidewall on each course to maintain a strong bond. The V-shaped gap at the rear will be filled with rubble. Brick ties every couple of courses hold the joints together. The brick ties in the concrete block will secure the brick front wall (or return). The small torpedo level will be used to level the back wall.

hearth, which on this job is 1 in. above the subfloor. Once the guide boards are marked, Hilley uses a guide string on mason's blocks to control the height and alignment of the brick courses as he lays them up.

Hilley picks sound, hard used brick for the firebox and hearth. The hearth is laid to the guide string in a good bed of mortar (drawing, top left). The firebox walls will be laid on this brickwork, so it extends beyond their eventual positions. Hilley likes to slope the hearth toward the back wall about ¼ in. to keep water from running into the room if any rain finds its way down the chimney. As with all brickwork, small joints look best, so pick your bricks for uniform thickness (see articles on pp. 78-81 and 82-85).

Laying out and building the firebox—With the hearth laid, Hilley finds the centerline of the opening, and marks off 2¼ bricks on each side for a 36-in. opening. Standard bricks are 8 in. long by 3¾ in. wide by 2¾ in. deep, but these measurements can vary, especially with used brick. Hilley uses bricks instead of a tape or ruler for an accurate layout, because 4½ used bricks (two times 2¼), laid end to end, don't always total exactly 36 in. The line of the back wall is 20 in. from the front line, and its length is figured by counting a little more than a brick on each side of the center line. Hilley pencils these lines on the brick hearth.

The lines for the diagonal sides of the firebox are drawn between the ends of the front and back lines. Where the side line meets the front line at the juncture of pilaster and firebox wall, you can draw either a mitered corner, or a square corner (drawing, bottom left). I like the look of the mitered corner, and I think the time it takes to cut the bricks is worth it. Cutting brick with a masonry blade in a skillsaw is easy when the brick is held securely between two cleats nailed to a plank. Both pieces of the cut brick are used, so cutting halfway through from each side is a better way to go.

One way to achieve a mitered look without cutting is to start a full brick at the front corner and butt the front return brick to the back corner of the starting brick. The triangular gap in front can be filled with mortar and covered with a tile facing, finish parging, stone, or the like, as shown in the drawing at left.

When Hilley is doing a square-cornered fireplace, he brings the side walls to a point ½ in. back of the edge of the return. This gives a neat line, which is very important with used brick because its width can vary from 3½ in. to 4 in.

Firebrick isn't required when you're building a firebox like this one, but Hilley uses it because heat-stressed common brick sometimes fractures violently. Most people don't like the look of firebrick in a Colonial fireplace, so he uses it only for the first six or eight courses—just high enough to cover the hot spot of a fire. You can see this blackened hot spot on the back wall of any fireplace. After a few fires, the firebricks soot up and blend in with the used brick in the rest of the fireplace. Hilley doesn't use refractory cement with the firebrick, but he does keep his mortar joints under ¼ in. thick.

Hilley begins by sprinkling sand or spreading a piece of building paper on the brick hearth. This simplifies cleanup later. Then he lays up four courses of the back wall plumb, level, and parallel to the front—a small brick wall about 20 in. wide by about 11 in. high. The fifth course is a tad longer. It's also tilted or rolled in slightly by troweling on more mortar at the rear of the joint than at the front. This is the beginning of the curved back wall (photo facing page).

Next, five courses of the mitered side and front wall are laid up using the angle-cut brick at the front corners and by cutting and butting the rear brick to the back wall. The way to do this at the back wall is to score each end brick in the back wall with the tip of the trowel as you hold the brick in the rolled position. The coving is plumb, so the trowel should come off the bricks of the coving below and follow through in a plumb line, as shown in the drawing below. The scratch is very visible, and cutting is done with a brick chisel or the sharp end of a mason's hammer.

The two pieces of angle-cut brick at each front corner should fit together tightly where they show, and the V-shaped gap behind should be filled with mortar and a piece of brick. Hilley also likes to use a brick tie across this corner every couple of courses. This corner can get out of plumb easily, so a constant check with a level is a must. If a running bond

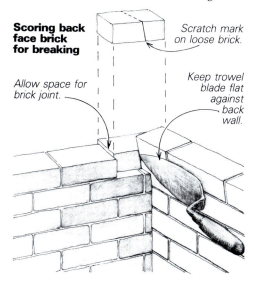

Scoring back face brick for breaking

Scratch mark on loose brick.

Allow space for brick joint.

Keep trowel blade flat against back wall.

is to show on the lintel course over the opening, you will have to watch the bond on your pilasters so that it will flow right into the bond on the lintel course.

Continue by rolling a few courses of the back wall, then building up the side walls. The roll will produce a gentle curve up to the damper, and it will make the back wall wider at lintel height than it is at the base. Each back-wall course is a little longer than the one below it, which is why the end bricks have to be marked in place for cutting. When a back-wall course needs to be a tad longer than two bricks, Hilley stretches it by setting a half-brick, or less, over the middle of the back

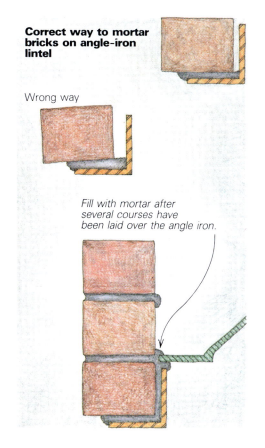

Correct way to mortar bricks on angle-iron lintel

Wrong way

Fill with mortar after several courses have been laid over the angle iron.

course below. The stretch, in other words, is accomplished in the middle of the course, not at its ends.

It is important while you're laying up the firebox to keep the side walls plumb. (In a square-cornered fireplace, the front and back walls are laid up first, a few courses at a time. The side walls are filled in.) You also must keep the back wall parallel with the hearth bricks. To do this, eyeball down the face of the back wall as it is laid, or measure from front to back on each side.

At the top of the firebox, the width of the opening from the outside face of the lintel brick to the rear face of the back-wall brick should be around 16 in. Hilley's formula for the amount of roll to give each back-wall course is simply experience. This is how most masons work. I'm always amazed at the way they seem to come out exactly where they want to be with exactly the right-sized opening, with no measuring at all. A novice might want to make a cardboard template to use as a guide, or spring a thin strip of wood against the first few courses to see how the curve projects up to lintel height.

Standard firebrick is thicker than used brick, so the back-wall courses will be higher than the side-wall courses. But the height should even out by the time you reach the lintel because the upper back-wall courses are tipped or rolled forward. As the back wall approaches lintel height, you can see how its courses relate to those of the front and side walls. By varying the joints, the wall heights can be adjusted to match.

When the firebox is at lintel height, Hilley fills in the space between the concrete-block wall and the back face of the firebox almost to the top with loose rubble. The rubble acts as a

The lintel. Side walls, back wall, and angle-iron lintel are at the same height to support the damper. The first course of bricks over the lintel overhangs the flange of the angle iron, and these bricks have to be laid up carefully so they won't roll forward. Pieces of building paper tucked at the ends of the angle iron serve as expansion joints.

heat sink, and more important, keeps the firebox positioned while allowing for expansion. A little mortar thrown in now and then will keep some of the rubble in place if a burned-out brick ever has to be replaced.

The lintel—A very important step in fireplace building is the proper installation of the angle-iron lintel. In this 36-in. fireplace, Hilley used 3-in. by 3-in. angle iron, which he installed with its ends bearing 1 in. or so on the pilaster bricks with a minimum of mortar—just enough underneath to stabilize it. The lintel's ends must be free to expand, and to ensure this Hilley tucks rolled-up scraps of building paper at each end. They act as spacers, keeping mortar and brick away from the angle-iron ends, and allow it to move.

The bricks in the first course above the lintel overhang the steel, and they have to be laid carefully (photo above) so that they won't roll forward. To help keep them from rolling, Hilley doesn't trowel any mortar behind them until a few courses have been laid, as shown in the drawing above. This eventual filling in, though, is important. Hilley feels that it prevents distortion of the angle iron from excess heat.

The damper—The damper should be sized to cover the firebox opening. The opening should be about as wide in front as the damper's flange, and from 2 in. to 5 in. narrower at the rear, depending on the damper's shape. The front flange rests on the top edge of the angle iron, and the side and back flanges rest on the firebox brick. The damper should be set in a thick bed of mortar on the brick and angle-iron edge, after three lintel courses are laid up, as shown in the photos at

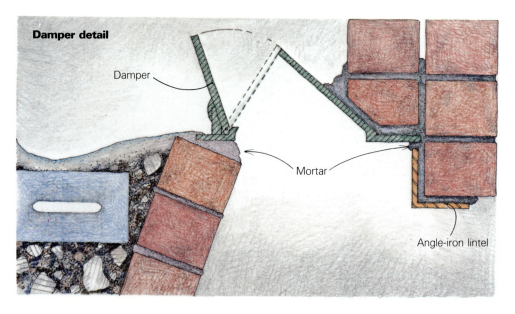

Damper detail

Damper

Mortar

Angle-iron lintel

The damper is mortared in place after three lintel courses are laid up. The space between the back wall of the firebox and the concrete-block core is ready for loose rubble fill, as shown in the drawing above.

The smoke shelf behind the damper is a 1-in. mortar cap over 4-in. concrete blocks on top of the loose rubble fill behind the firebox. The damper side is higher than the rear so any rainwater will drain away from the opening.

Laying up the smoke chamber is not fussy work. Hilley uses soft brick and concrete block, and then he parges the smoke shelf and chamber walls with mortar.

center left. As with the angle-iron lintel, it is important to keep masonry away from the ends of the metal to allow for expansion.

Smoke chamber—The smoke chamber is the open area behind the damper, where cold air coming down the chimney bounces off the smoke shelf at the bottom and is deflected upward, along with smoke rising from the firebox. As a base for the smoke shelf, Hilley lays a flat course of 4-in. concrete block on top of the rubble and concrete-block back wall. He sometimes lays a few concrete blocks, dry, directly on top of the loose rubble behind the rear wall. Then about 1 in. of mortar is smoothed out to make the smoke shelf's surface. Rainwater will puddle up here, so pitch the shelf away from the firebox and trowel it well. (Accumulated water will eventually evaporate or be absorbed into the masonry.)

The smoke chamber (drawing, p. 95) is formed by rolling the bricks of each course inward until the opening at the top is the size of the chimney flue tile. Hilley rolls the bricks a few courses at a time, alternating the corner bricks to maintain a bond.

Where the rolled brick courses meet at a corner, Hilley breaks off a piece of the lead corner for a better fit. He uses soft, spalling used bricks for this work. They are easy to shape, and it's not fussy work. In fact, Hilley had me hold up a sagging wall while he finished an adjacent supporting corner. A wall will collapse if laid up too much at one time.

Hilley says rolling the bricks to meet an 8x10 flue should give you a smoke chamber 24 in. to 36 in. high. Don't reduce from damper size to flue size too fast, and keep the smoke chamber symmetrical. Hilley once built a fireplace with the flue on the right side of the smoke chamber. This created unbalanced air pressures in the chamber and caused little puffs of smoke on the right side of the firebox.

The inside face of the smoke chamber is parged with mortar. (Be sure you leave enough clearance for the damper to open.) A piece of building paper or an empty cement bag laid on the damper before parging will keep things clean. You don't want your damper lid locked in solid with mortar droppings. The smoke-chamber walls must be 8 in. thick, so Hilley builds out their lower part with interlocking brickwork, and the upper part with flat-laid 4-in. concrete block. Then he parges the whole business with a layer of mortar (photo bottom left).

The rolled brick and outer block shell of the smoke chamber transfer the flue and chimney weight to the lintel, keeping the lintel bricks in compression. The first flue tile sits on top of the smoke chamber, fully supported by the brick, and the chimney is built around it. Brickwork against a flue will crack as the hot flue expands, so there must be at least a 1-in. airspace between the tile and the chimney shell. If the chimney is concealed, the masonry can be concrete block. □

Consulting editor Bob Syvanen is a carpenter in Brewster, Mass. Photos by the author.

Laying Brick Arches

A masonry inglenook becomes the warm center of an architect's new house

by Elizabeth Holland

When Ed Allen and his wife Mary started thinking about designing their own house seven years ago, they knew they wanted something big and barnlike, yet New England simple and cozy. Ed, an architect, had been fascinated by domes, vaults and arches during his studies, but it wasn't until the couple spent six months living in Liverpool during the winter of 1975-1976 that an inglenook became part of their house plans.

"We spent a lot of cold evenings huddled around a fire, trying to keep warm in an unheatable English house," remembers Allen. "We kept sketching our ideas and dreaming about how we wanted to have a house where we wouldn't be so cold."

The sketches showed the influences of the English and Welsh houses they had seen, particularly of their visit to the Welsh Folk Museum at St. Fagan's. This is where the Allens encountered the inglenook.

It is thought that the word *ingle* comes from the Gaelic *aingeal*, meaning fire or light, and was originally applied to open fires burning on primitive hearths. In medieval times, it came to mean a fireplace. The inglenook was a corner or a small room near the chimney where the family would gather before the heat of the flames. This became the central idea in Ed Allen's plans, and the inglenook ultimately became the dominant design element of the house he was to build.

Allen spent a long time working out the exact dimensions of the inglenook, and was working on a 4-in. module. Then the house was designed around that, from the inside out. "I was always trying to keep it as small as I could," Ed says, "but it kept growing as I made sure all the spaces around it were the proper size."

The Allens' inglenook is all brick, part of a 50-ton masonry mass that encompasses flues and fireplaces, and divides the house into two equal parts on each floor. The inglenook was designed to accommodate four or five people comfortably. The interior dimensions are 7 ft. 4 in. wide and 7 ft. deep from front to back, not including the fireplace.

According to Allen, it was all worked out logically—the dimensions of what he wanted things to be, plus the dimensions of the necessary brick. For example, the archway is 16 in. thick because the walls contain an 8-in.-square flue plus 4 in. of brick on each side.

To make the layout easier, Allen designed the brickwork to be modular, with 7⅝-in. long bricks and ⅜-in. mortar joints, for an overall length of 8 in. Later he discovered that only about two-thirds of the assorted Full Range Belgian bricks he ordered were the right size. The rest were up to ¼ in. too long. This required some cutting when he got to his closers, the last bricks in each course (for tips on cutting brick, see pp. 73, 85, and 105).

Mortar color and tooling can significantly affect the look of brickwork. Allen chose a standard dark masonry cement. On the horizontal mortar joints, he used a flat-joint finishing tool to make a weathered joint, flush with the brick at the bottom and cut back at the top. This joint casts a shadow on the mortar joint, accentuating the pattern of the brick. The vertical joints, however, are gently concaved.

Choosing a bond—Bonding, the overlapping patterning of the bricks, knits the various wythes (thicknesses of brick) together. For a single wythe, a simple running bond can be used. But structural brickwork is usually at least 8 in., or two wythes, thick. "An 8-in. thick wall can use any of a variety of bonds, and some of them are quite beautiful," says Allen. "I was planning to use an English Garden Wall bond because if you're doing an 8-in. wall where one side is going to be concealed, you can save a lot of time by laying up the concealed wythe in 4-in. concrete blocks."

An English Garden Wall bond consists of three courses of stretchers (bricks with a long edge showing) and a fourth course of headers (bricks with their short ends facing out and

---

Bricklaying tips for amateurs

After years of studying and teaching brickwork, and laying bricks, Ed Allen is convinced that interior brickwork—including arches—is well within the grasp of a reasonably careful amateur.

"Once you get into masonry, it's just like putting up a wood frame for a house—absolutely routine, very secure, very simple," he explains. "There's just no end to what you can do with it."

Bricklaying is a relatively straightforward concept that gets fairly complex in practice. Here are some tips for beginners:

- You can't learn brickwork on your own. Learn from someone who knows how to do it. You can pick up the rudiments in a day or two. "Probably 90% of the success in laying brick is getting the mortar the right consistency and using the trowel properly. Learning to mix mortar to the proper consistency and to use a trowel are things that simply can't be gotten from a book."

- Plan in advance and know your dimensions. Determine the heights at which the arches will spring, and the heights of the arches. Consider the placement of flues when determining thickness.

- An arch supports a vertical load by transforming it into a diagonal load. Make sure your design has enough mass on either side of the arch to absorb the thrust and keep the arch from spreading.

- The labor involved in laying brick is almost directly proportional to the number of corners. Called leads, the corners are laid up first, four to six courses at a time. Care in laying up the leads pays off in level courses of the right height. The bricks that fill in the flat stretches between corners, aligned with strings pulled taut between the leads, are laid relatively quickly. It's best to eliminate as many corners as possible. You've always got to decide, of course, whether it's worth the extra labor to get it the way you want it.

- Practice by building the foundations for whatever you are going to build. This gives you a chance to develop techniques before laying up courses that show. If you're still having trouble, get some help.

- It's easiest to work at waist level, or roughly between your knees and shoulders. Arrange the scaffolding accordingly. As you go higher, your work slows down because it becomes more cumbersome to transport heavy and bulky materials.

- Brickwork is not as precise as carpentry. The irregularity of it is part of the charm, and a real plus for amateurs. Nevertheless, you should strive for precision, so things don't get too far out of whack. —*E.H.*

their lengths extending in across the two wythes). On the concealed wall, a course of concrete block takes the place of three courses of brick plus their mortar joints. The fourth course is composed of the headers, lying across both the brick and block wythes.

A header course is laid so the bricks straddle the vertical mortar joints in the stretcher course below. Since Full Range Belgian bricks are narrow, spacing the bricks properly would have required extra-wide joints. This led Allen to change from a full header course to what's called a Flemish header course (drawing, right), where headers and stretchers alternate.

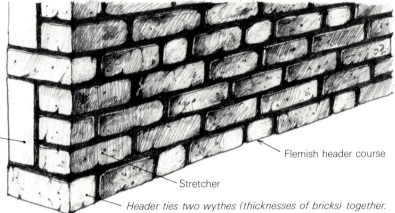

Concrete block is 4 in. thick and as high as three courses of brick with mortar joints. Using block where it won't show saves time and money.

Flemish header course

Stretcher

Brick bonding

Header ties two wythes (thicknesses of bricks) together.

Building the arch.
Allen and a friend did all the masonry work themselves, in two stages. First, in May, they laid up the block foundation for the masonry core. The brickwork began in July, and from then until mid-autumn they laid 7,000 face bricks, an undetermined number of concrete blocks and concrete bricks, and several hundred sections of flue tile. The graceful curve of the inglenook's archway is inviting, a welcome contrast to the sturdy straight lines of the brick walls. Within the inglenook and overhead, the arch repeats, each time in a slightly different form.

"Arches are really fun—they're a wonderful structural form," Allen reflects. "I think everyone has an immediate, positive emotional response to them. By the time we finished the arches we decided they were quite easy to build and not terribly time-consuming, a conclusion quite contrary to our initial expectations."

Step one: the centering form—The brick walls are laid up to the level called the springing of the arch, the point where its curvature begins. Now the centering is built—the wooden form over which the bricks in the arch will be laid.

There are many ways to build a centering. Mine consists of two identical curved trusses with a single rectangular piece of ¼-in. Masonite nailed securely to the top of them (drawing **A**, below). You should check the centering's fit by holding it between the brick sidewalls.

Step two: patterning the arch—Lay the centering on its side on the floor. Stand the bricks you're going to use on end, all around the curve (**B**). You should have an odd number of them. Using an even number results in a mortar joint positioned at the crown of the arch. This looks bad and makes the structure weak. Because of the curvature of the arch, the mortar joints will be wedge-shaped. They should be about ³⁄₁₆ in. wide at their narrowest point, next to the centering. Shuffle the bricks around until the spacing looks good. You could have wedge-shaped bricks specially made, but these are usually expensive, and must be ordered well in advance of when you'll need them.

With a pencil, mark the thicknesses of the mortar joints on the centering itself. Take away the bricks and use a square to run the joint lines across the curved surface of the form. When you're finished, the centering will be marked to show you where all the bricks belong, and how thick the mortar joints should be. This step is crucial—without it you will end up with uneven joints, and you'll have to trim bricks to fit odd spaces.

Step three: placing the centering—Lift the centering between the existing brick walls. The bottom of the centering should be just a couple of inches below the spring of the arch (**C**). Support it by four lengths of 2x4, cut just a little longer than the distance from the floor to the spring of the arch so they can be angled under the four corners of the centering and wedged in place. It's best to align the centering so its front edge is even with the brick walls; then bricks in the arch can be laid to the front edge of the form. If your centering is wider than the arch, pencil a line on it to indicate where the front ends of the arch bricks should be laid. Level up by tapping the bottoms of the wedged 2x4s. Once the centering is level and in the right position, drive shims into the small gaps between the centering and the walls at the four corners, to hold the form firmly in place.

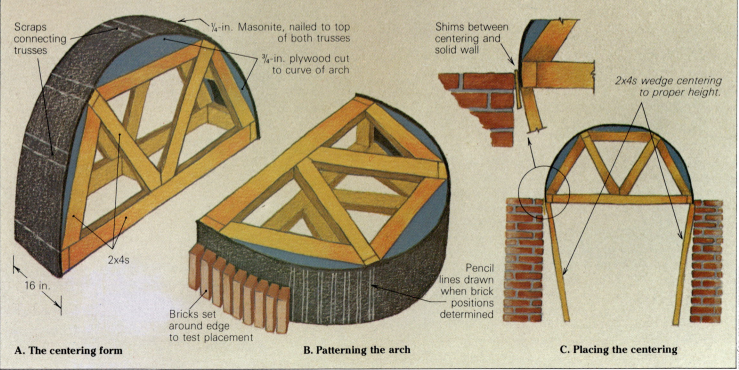

Scraps connecting trusses

¼-in. Masonite, nailed to top of both trusses

¾-in. plywood cut to curve of arch

2x4s

16 in.

Bricks set around edge to test placement

Pencil lines drawn when brick positions determined

Shims between centering and solid wall

2x4s wedge centering to proper height.

A. The centering form

B. Patterning the arch

C. Placing the centering

Illustrations: Frances Boynton

D. The ends of the bricks in the arch must be in the same plane as the wall. Check for alignment with a trammel board tacked to the center of the horizontal truss member.

E. Bricks are set between the pencil lines that were drawn on the centering during a test-fitting on the ground. This guarantees a proper fit and mortar joints of uniform size.

F. Bricks have been trimmed with a mason's hammer where wall meets arch. It looks best if the width of the curving joint between wall and arch remains constant.

Step four: laying up the arch—Lay bricks from both lower edges until they meet at the top. As they go up, check the bricks with a level to make sure their ends are in the same plane as the wall, or with a board tacked to the center of the truss (**D**). If you follow your pencil marks, you'll end up with the right number of bricks, and uniformly wide joint spaces (**E**).

Step five: finishing the wall—With the centering still in place, lay up the flat walls around the sprung portion of the arch. If the centering is removed too soon, the arch might collapse, because the mortar is still fresh and because a semicircle is not the strongest arch form. It could well bulge out at the sides and drop in at the center. Once the walls are laid up around the arch, however, it can support a lot of weight.

In the Allens' house the vaulted ceiling over the inglenook carries a concrete slab floor above it.

Always work in from the leads at the outer end of the wall, keep the vertical joints lined up properly and use taut string lines to keep the courses perfectly level.

On each course you will have to cut the last brick to fit, where it intersects with the arch. A diamond saw is the most efficient tool for this, very precise and sure. But a diamond saw is expensive and to Allen's eye the cut is too sharp-edged and cold. An abrasive masonry blade ($5 to $6 in a hardware store) in a circular saw is a good alternative. "It's a messy operation and it doesn't cut the brick as well as a diamond saw," he says, "but you can score the brick deeply and then crack it with a hammer to get a pretty clean break."

For his own arches, Allen used a mason's hammer to cut the bricks. This takes some skill, he cautions, and results in a somewhat ragged break and a lot of wasted bricks because you can't always get the bricks to break exactly as you want. The flatter angles that are required near the top of the arch are particularly difficult to make with a hammer.

A ragged cut on a few bricks, though, is less important than making sure that the curved mortar joint between the arch and the entire brick wall around it is a constant thickness (**F**). A curved joint that varies in thickness is unattractive, and once it's there, you can't do much about it.

Once the courses are laid up around the arch, tap out the 2x4s carefully, drop the centering gradually, then remove it and admire your arch.

Tooling the joints—In typical brickwork, after the brick is laid in the wet mortar, the mason cuts off the excess mortar at the face of the brick. In one to three hours the mortar will be thumb-print hard, the proper consistency for tooling with a V-shaped or rounded metal rod called a jointer or striking tool. This produces a clean and attractive joint. (For exterior brickwork, tooling is doubly important because it helps compact the mortar at the face, making it much more weather-resistant.)

But with an arch, the bottom mortar joints can't be tooled when the mortar is thumb-print hard, because the centering is in the way. By the time it is removed, the mortar is too hard for tooling.

Yet old arches and vaults have well-tooled joints. Allen was puzzled. Books on the subject were no help. Several masons suggested rubbing the joints with full-strength muriatic acid. The acid quickly dissolved the excess mortar that had been stuck between the brick and the centering, but the joint remained undefined and fuzzy.

Allen eventually learned that before the era of portland cement, arches and vaults had been laid up with lime mortar, which sets very slowly. He says that even if the centering were kept in place for several weeks, when it was removed the lime mortar would still be soft enough to be tooled. Although lime mortar is not as strong as portland cement, the arches themselves are structurally stable enough to be able to compensate for the weaker mortar. Allen recommends either using lime mortar alone for the arches, or using it for only the bottom part of the joint that meets the centering, and then using mortar made with portland cement for the work above this.

Living with the inglenook—The Allens' inglenook can hold up to six people within its candlelit confines. The snug spot sits off a spacious country kitchen, and is used more often on social occasions than on weekday evenings. "It can get a little crowded," Allen says, " but in that kind of space it doesn't feel crowded. I think people are accustomed to drawing close around a fire." □

Elizabeth Holland lives in West Shokan, N.Y. She writes about the design and construction of energy-efficient buildings, and is contributing writer for this magazine.

Rumfordizing Brick by Brick

How to convert an energy-wasting fireplace to an efficient heater

by Kent Burdett

Most modern fireplaces don't do a very good job. Many smoke so badly that they can't be used, and almost none are efficient heaters. In fact, many of them draw more heat out of the house than they return, sucking in warm room air and sending it up the chimney. But it's possible to convert one of these mere ornaments into a functioning and efficient fireplace. An American Tory named Benjamin Thompson, later called Count Rumford, demonstrated the relevant principles two centuries ago.

Rumford proved that the key to an efficient fireplace is a properly proportioned firebox, with important dimensions based on the width of the opening. (The parts and proportions of a Rumford fireplace are shown on the facing page.) Both the firebox's depth (distance from opening to fireback) and the width of the fireback should each equal one-third the opening's width. This makes for a shallow firebox with covings angled at a sharp 45° to reflect the fire's radiant heat into the room. The fireback, which must be vertical to a height equal to one-third the opening's width, begins to slope forward from that point to a small throat above the lintel. The sloping back reflects more heat, and the small throat results in a more forceful movement of air up the chimney. It also leaves room for the smokeshelf, a necessary feature where descending cool air and ascending hot air circulate to set up a strong draft.

Rumford's workmen renovated so many smoking fuelwasters in England that a new word entered the language. His wealthy customers didn't just have their fireplaces improved, they had them "rumfordized." Once you know the principles, you can rumfordize your own fireplace. Fireboxes are not structurally connected to the masonry of the chimney, so you can tear out an unsatisfactory one and replace it easily.

Materials—For this job you'll need clean sand and water, masonry cement for the rubble fill behind the new firebox, firebricks and fireclay to join them. Experience provides the best way to judge just how much of each you will need for a specific project. For a fireplace 24 in. wide and 30 in. tall, I used 84 firebricks (2¼ in. by 4½ in. by 9 in. each), along with ¼ yard of sand and 2 sacks of cement. You can buy clean sand by the fraction of a yard at most lumberyards.

There are two kinds of fireclay, premixed and dry. The premixed costs twenty times as much as the dry variety, and can't be scraped or chipped from the faces of firebricks after it dries. Dry fireclay (available at ceramic supply houses) is sold in 50-lb. sacks, but one sack costs less than a single gallon of premixed.

As for the firebricks, try to buy what you need from the same lot. They will be fired to the same hardness, and there will be fewer small variations in their dimensions. You can find them at masonry supply houses and many lumberyards.

You probably already own or have access to most of the tools you'll need: a lightweight hammer, a tape measure, a try square, a level, a soft

---

Benjamin Thompson, Count Rumford
by Simon Watts

An efficient fireplace was hardly Benjamin Thompson's only contribution to civilized living. He was an extraordinary American whose adventurous life and considerable scientific achievements remain largely unknown. He was an ingenious inventor, always trying to improve clothing, coffee pots, eating habits, lamps and whatever else crossed his path. Perhaps he was overshadowed by his great contemporary, Benjamin Franklin, but I suspect his obscurity has more to do with his having been on the wrong side of the American Revolution.

Born in 1753, Thompson showed his scientific bent early. While still in his teens he experimented with gunpowder and electricity, and was already keeping a detailed journal of his observations, which became a lifelong habit. On at least one occasion his scientific curiosity nearly cost him his life. Attempting to repeat Franklin's famous experiment, he constructed a 4-ft. kite and flew it in a thunderstorm. Going the more prudent Franklin one better, he soaked the kite string in water to make it a better conductor. The results were suitably dramatic: Watching from the house, his family was amazed to see the youthful experimenter outlined in fire. He later remarked in his diary, "It had no other effect on me than a general weakness in my joints and limbs and a kind of listless feeling. However, it was sufficient to discourage me from any further attempts."

In 1772 Thompson was invited to teach school in Concord, N.H. Within a few months he married a wealthy young widow, and for the first time was financially independent. He settled down to manage his wife's estates and pursue his scientific studies, but those were restless times. Rebellion was in the air, and the colonists were taking sides. Thompson remained loyal to the king, barely escaped being tarred and feathered, and fled to England, where he was put in charge of recruiting, equipping and transporting British forces in North America.

After the war he went to Bavaria, where he was given the job of reorganizing the Elector's woe-begone army. With characteristic thoroughness, Thompson spent several years making a detailed study of the army, and finally came up with a plan so comprehensive that it stunned his critics speechless. Thompson's report focused on the army's two major expenses—food and clothing. Questioning the existing cloth, he set about experimenting with different materials, such as fur, feathers, cotton, wool and jute, to find out which were the cheapest and most effective for soldiers' uniforms. He devised ingenious experiments to compare thermal conductivity, and was the first to suggest that it was not the material, but the air trapped in the fibers, that provided insulation. He then designed a new cloth and looked around for a firm to weave it. None of the existing companies was willing to cooperate, and they all thwarted his every attempt to set up a new factory.

This setback prompted Thompson to make his most spectacular experiment in social reform. At that time Munich was plagued by professional beggars so numerous and well organized that they practically ran the city, even intimidating the police. Thompson saw an opportunity both to staff his factory and to rid the city of its beggars. On New Year's Day, 1790, he made his move, and before nightfall every beggar had been arrested and locked up in what was euphemistically called "The Poor People's Institute," but which was actually a workhouse. Thompson ran the Institute with a firm hand, and within a few months the former beggars had been trained to produce cloth of an acceptable standard.

By 1791 Thompson had become a general in the Bavarian Army, as well as Minister of War and Minister of Police. In 1792 he was made a Count of the Holy Roman Empire, and adopted the name of Rumford—the original name for Concord, N.H.

If you're interested in reading more about this singular man, try Sanborn G. Brown's *Count Rumford: Physicist Extraordinary* ($18.25 from the Greenwood Press, P.O. Box 5007, Westport, Conn. 06881).

Simon Watts is a writer and cabinetmaker in Putney, Vt.

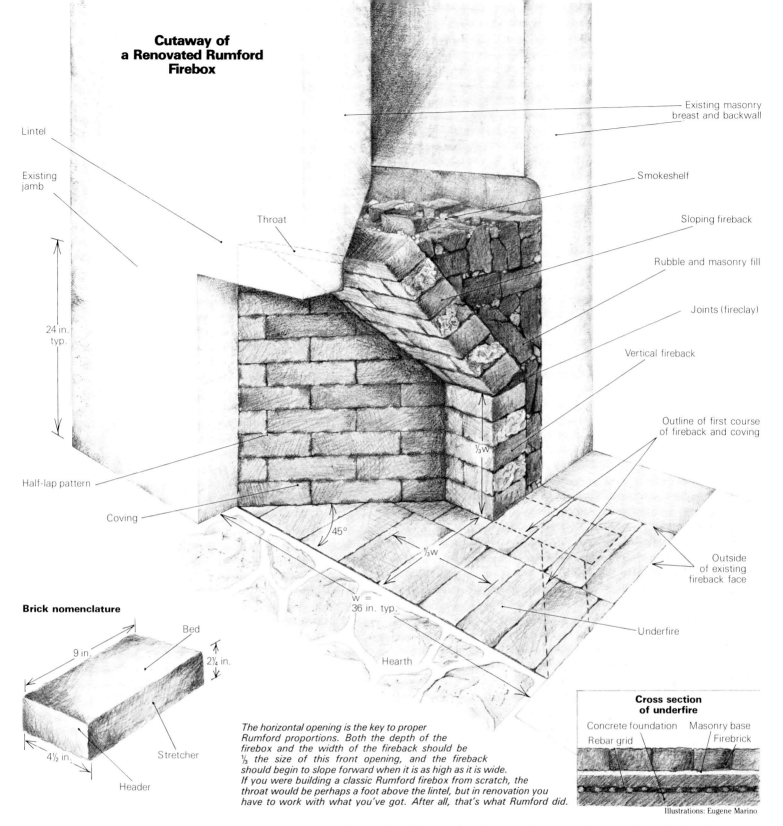

Cutaway of a Renovated Rumford Firebox

Lintel

Existing jamb

24 in. typ.

Half-lap pattern

Coving

Throat

Existing masonry breast and backwall

Smokeshelf

Sloping fireback

Rubble and masonry fill

Joints (fireclay)

Vertical fireback

⅓ W

Outline of first course of fireback and coving

45°

⅓ W

Outside of existing fireback face

w = 36 in. typ.

Underfire

Hearth

Brick nomenclature

Bed

9 in.

2¼ in.

4½ in.

Stretcher

Header

The horizontal opening is the key to proper Rumford proportions. Both the depth of the firebox and the width of the fireback should be ⅓ the size of this front opening, and the fireback should begin to slope forward when it is as high as it is wide. If you were building a classic Rumford firebox from scratch, the throat would be perhaps a foot above the lintel, but in renovation you have to work with what you've got. After all, that's what Rumford did.

Cross section of underfire

Concrete foundation   Masonry base

Rebar grid   Firebrick

Illustrations: Eugene Marino

brush, a string, a shovel, a hoe and a wheelbarrow for mixing cement. You will also need a 4-in. brick set (a chisel for breaking bricks), a 12-in. mason's trowel and a cold chisel or all-purpose masonry chisel. All these tools can be bought at most hardware stores.

Underfire—Often, all you have to do to remove an original firebox is to reach up, grab the top back brick, and pull (photo next page, top left). If the bricks don't tumble right down, use a hammer and an all-purpose chisel. This is a dirty job that stirs up a lot of dust, so be sure either to cover the furniture or to move it out of the room, and seal the doors to the rest of the house with tape. Save both the old firebricks and the rubble

behind them. You may be able to use them later. If there is already a damper mechanism at the throat, leave it. Cement from the chimney bricks or tile will hold it in place.

After pulling down the old fireback, covings and rubble fill, remove the bricks of the existing underfire. Using a chisel and hammer, firmly tap the masonry foundation beneath. If it seems solid, you can go ahead and lay the new underfire. If the foundation is cracked, loose or crumbling, however, you will have to replace it. Chisel it out to a depth of about 4 in., then lay a grid of ½-in. rebar 4 in. on center before pouring a new foundation of concrete. You will want the finished underfire to be level with the outer hearth, so take careful depth measurements and

leave enough room above the foundation pour for a base and the underfire brick. If a sound foundation is too high for the bricks you are using, you will have to chisel it out. When the foundation is ready, pour a base ¼ in. to ½ in. deep for the new underfire. Mix two parts of clean, dry sand with one part of masonry cement, then add enough water to create a pourable mixture. Spread it over the foundation.

Lay the firebricks for the underfire with no fireclay between them. The hearth takes a lot of abuse, and cracked brick can be broken away from the base cement and replaced easily if it has been installed this way.

Lay the first row of bricks beginning with the ones on the extreme right and left of the open-

Fireboxes aren't structurally connected to the masonry of the chimney. You can often tear one out, as at left, by reaching up and tugging. Above, the underfire should be installed dry, without fireclay between the bricks. Be sure each brick is set level. (The bottle is covering an old gas line that will become a fresh-air intake.)

When the fireback is only one brick long, the most efficient way to butt coving and fireback is to cut the coving brick to fit along the fireback's header, left. When the fireback is longer, cut the coving brick to fit against fireback stretchers. Above, firebox is filled in behind with rubble and masonry. A fairly wet concrete mix will flow to the bottom.

ing. Once these are in place, you can rest your level across them to be sure they and subsequent bricks are perfectly horizontal. Next, use a plumb bob to find the opening midpoint, and center a carefully leveled brick on it. Place and level bricks alternately on either side of the central one until you reach the two at the edges of the opening. The fit will probably be less than perfect, and these outside bricks will have to be trimmed. They will eventually be covered by the masonry of the new firebox, so all the bricks that show will be the same size. This looks good, and it also makes them easier to replace, if that ever becomes necessary.

Once the first row is set in place, complete the rest of the underfire. You don't need to cover the entire masonry base you've poured, just enough of it to provide a solid surface on which to build the fireback and covings. The rest will be cov-ered with rubble and masonry as you build up the firebox and fill in behind.

When the underfire is laid, draw in the lines of the fireback and covings. This is when the critical psychological problem arises. If you're not used to Rumford dimensions, the outline of your new firebox will look too shallow. You'll wonder if wood will fit, and whether such a fire-place could possibly draw well. Don't worry. A Rumford can be up to two-thirds more efficient than a squat, deep, modern fireplace. But be pre-pared for kibitzers telling you it won't work. This is such a predictable nuisance that I prefer to do this part of the job without an audience.

Fireback and covings—The night before you plan to lay the firebricks along the lines you've drawn, mix 2½ to 3 gal. of dry fireclay with enough warm water so that when a brick is dipped in, ¹⁄₁₆ in. to ⅛ in. of the mixture will ad-here to it. The next day, your technique will be to dip each brick's bed and headers into the fire-clay and lay up the covings and fireback with joints ¹⁄₁₆ to ¹⁄₃₂ in. wide.

Begin by laying up the first course of the fire-brick. You will build up the firebox using a half-lap pattern. The fireplace in the photos above has a front opening just 27 in. wide, so I used a single 9-in. firebrick to form the vertical fire-back. Firebacks are usually more than one brick long. For a half-lap pattern, start by placing a brick on each side of the midpoint of the line you've drawn on the underfire. Then lay bricks along the line to a point at least half a brick length beyond its end. The next course will start with a brick centered above the midpoint line, laid up so that each brick overlaps half of each of the two bricks beneath it. It's a good idea to

Marking and Breaking Brick

For a simple straight cut, mark the brick as shown in figure A, and lay it on a cloth sack or cement bag filled with clean, screened sand.

Hold the chisel as shown, and give it a tap just heavy enough to score the brick along one of the lines you've drawn — don't attempt to break it with one blow. Score

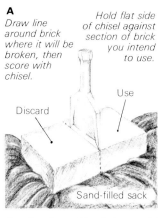

A
Draw line around brick where it will be broken, then score with chisel.

Hold flat side of chisel against section of brick you intend to use.

Use

Discard

Sand-filled sack

the other three sides, being careful to hold the chisel firmly against the brick, so it doesn't pop up after you strike it. The brick will break along the proper plane.

For the angled cuts required when the lower courses of firebrick and coving meet, the breaking technique is the same, but scribing

the lines to score along is somewhat more complicated (see figures B and C).

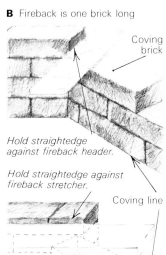

B Fireback is one brick long

Coving brick

Hold straightedge against fireback header.

Hold straightedge against fireback stretcher.

Coving line

Fireback

C Fireback is more than one brick long

Once the fireback begins sloping, scribing lines becomes a two-step process with the straightedge

because the angle must be cut through two planes (see figure D).

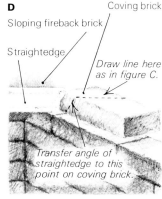

D
Coving brick
Sloping fireback brick
Straightedge
Draw line here as in figure C.

Transfer angle of straightedge to this point on coving brick.

To draw lines on the third and fourth sides, set your straightedge beneath the brick, align it parallel

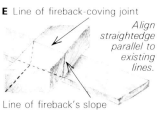

E Line of fireback-coving joint
Align straightedge parallel to existing lines.

Line of fireback's slope

to the existing lines as shown in figure E, mark the edges, and connect the dots. Break the brick the same way as for other angles, striking the brick perpendicularly. — K.B.

Establishing the sloping fireback

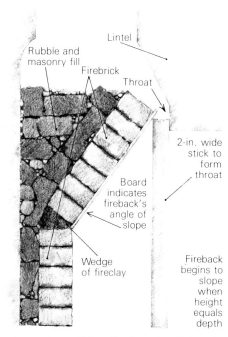

Lintel
Rubble and masonry fill
Firebrick
Throat
2-in. wide stick to form throat
Board indicates fireback's angle of slope
Wedge of fireclay
Fireback begins to slope when height equals depth

The throat of an efficient fireplace should be only 2 in. deep. Its depth and location determine the slope of the fireback. Use a stick to form the throat, and lean a board against it from the top of the vertical fireback. Then lean a firebrick against the board and fill in with a wedge of fireclay.

---

pre-mark the center of each brick, so that they can be set quickly in place. At either end, use a half brick, so that the courses come out the same length. Alternate these methods as you build the fireback up course by course.

With a single-brick fireback, the simplest place to join coving to fireback is along the fireback bricks' headers (photo facing page, center left). On longer firebacks, fit the deepest coving brick against the stretchers of the extended fireback bricks. Both of these techniques require cutting coving bricks at angles. Building up the firebox will require a number of such angled cuts, and even, as the fireback begins to slope forward, double-angle cuts. The drawings above show how to deal with this, probably the most technically difficult part of building a firebox.

After this brick is cut to the proper angle, set it aside, and begin laying bricks from the front. Set more bricks along the coving line until you're close enough to the fireback to bridge the gap with the brick you've set aside. Transfer the measurement of that distance to the short stretcher of the brick, cut it to length, and lay it in place.

The second course of the coving should begin with a half brick at the front to achieve the half-lap pattern, but the rest of the procedure is the same. Right and left covings should be mirror images of each other.

As you build up the fireback and covings, fill in behind with rubble and concrete (photo facing page, center right). You may be able to use much of the rubble you pulled out when you tore down the original firebox. You can also use broken brick you know you won't need to build up the new firebox. I usually use a concrete mixture of three parts sand to one part cement, though I often use a two-to-one mixture if it's already

around and handy. Add enough water to achieve a fairly wet consistency, so the concrete will flow to the bottom of the rubble.

Throat—When the fireback is as high as the firebox is deep, it's time to start sloping it forward. It is at this point that you have to decide how wide a throat your fireplace should have, because this will determine the angle of the slope. Almost all modern fireplaces have much too large an opening at the throat. Vrest Orton, in his book *The Forgotten Art of Building a Good Fireplace* ($3.95 from Yankee, Inc., Depot Square, Peterborough, N.H. 03458), says that the opening should be just 4 in. deep. I believe that even this is too much for anything smaller than a fireplace of a baronial hall. For most installations, 2 in. is more suitable.

Exactly where behind the breast will the throat fall? This is where the difference between building a fireplace from scratch and renovating one in the space allowed becomes most evident. In a classic Rumford, the throat will be perhaps a foot above the lintel. However, you probably won't have room to work much higher than a few inches above the lintel, so you'll have to form the throat there.

Begin by finding a stick of wood with a 2-in. dimension, and stand it vertically in the front of the firebox to mark the width of the throat. Remember that a milled 2x4 isn't the right size in any plane, but you can rip it to a true 2 in. along its nominal 4-in. face. Next, cut a flat board, perhaps a piece of plywood, just long enough to extend at an angle from the front of the fireback's highest course to the vertical stick at the point you want to form the throat.

This board describes the fireback's slope. Take a firebrick and, keeping in mind the half-lap pat-

tern you want to continue, reach behind the board and set the brick's front stretcher against the board's back. Fill beneath the brick with a wedge of fireclay, let it set up for 10 to 15 minutes, then take down the board and remove the vertical stick. Lay up the rest of the course, using your level as a straightedge to make sure that the rest of the bricks slope at the same angle, and allowing each wedge to set up. Once the slope is formed, subsequent joints on the fireback need no time to cure. Build the covings up course by course with the fireback. The fireback widens as it slopes forward, so you will have to extend each course a half brick or so beyond each end of the previous one.

As the fireback widens the covings get shorter, but the fireback's slope means that the final coving brick on each course must be cut at angles through two planes rather than just one.

If there was no original damper, you'll have to insert a new one before you build the firebox too high. Damper mechanisms can be bought at many lumberyards and brickyards. Get one to fit the width (not the 2-in. depth) of your fireplace's throat. You can wedge it up out of your way as you work to finish the covings and fireback. The rubble and masonry fill behind the fireback will create a flat smokeshelf on which the damper will eventually rest. You needn't fasten the mechanism down; just fill with concrete any gaps between metal and masonry.

You may light a fire in the firebox at any time during construction with no ill effects. Remember, though, that you wouldn't race your car until it's warmed up. Build a small fire at first, and only gradually stoke it up to a roar. □

Kent Burdett is an Oklahoman who has been installing and renovating fireplaces for 15 years.

The Fireplace Chimney
Flashing and capping are the tricky parts of the job

by Bob Syvanen

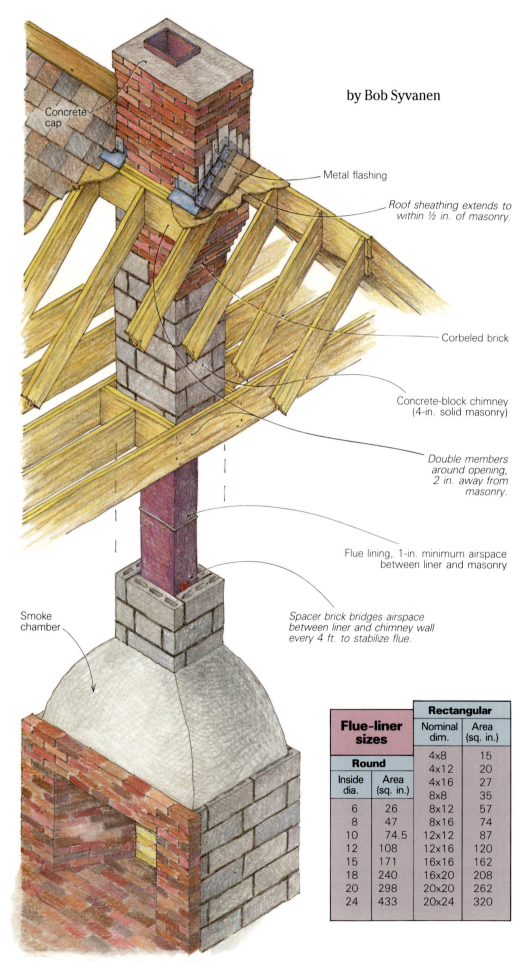

Concrete cap

Metal flashing

Roof sheathing extends to within ½ in. of masonry.

Corbeled brick

Concrete-block chimney (4-in. solid masonry)

Double members around opening, 2 in. away from masonry.

Flue lining, 1-in. minimum airspace between liner and masonry

Smoke chamber

Spacer brick bridges airspace between liner and chimney wall every 4 ft. to stabilize flue.

Flue-liner sizes		Rectangular	
		Nominal dim.	Area (sq. in.)
Round		4x8	15
		4x12	20
Inside dia.	Area (sq. in.)	4x16	27
		8x8	35
6	26	8x12	57
8	47	8x16	74
10	74.5	12x12	87
12	108	12x16	120
15	171	16x16	162
18	240	16x20	208
20	298	20x20	262
24	433	20x24	320

Throughout most of history, masonry chimneys were just hollow, vertical conduits of brick or stone. These single-wall chimneys have many problems: they conduct heat to the building's structure; they aren't insulated from the cold outside air, and so allow severe creosote buildup as the cooling gases condense; and they suffer from expansion and contraction, which lead to leakage of water (at the juncture of chimney and roof) and smoke (through cracks in the masonry).

In present-day chimneys, ceramic flue tile carries the smoke. It can withstand very high temperatures without breaking down, and also presents a much smoother and more uniform passageway, which means easier cleaning and thus less chance of chimney fires. Brick or concrete block, laid up around the tile but not in contact with it, serve as a protective and insulative layer.

Ceramic flue tiles can be either circular or rectangular in section, and are available in a number of sizes (see the chart below left). Brick, block and mortar are the only other materials you need to build a chimney (to find out how to build a fireplace, see the article on pp. 94-98). John Hilley, a mason I've worked with for the last eight or nine years, uses type S mortar throughout the entire chimney, but some codes require refractory cement to be used for all flue-tile joints.

Requirements and guidelines—Before you start building a chimney, you've got to consider sizing, location and building-code requirements. I'll be talking about masonry chimneys that are built above fireplaces, but most of the construction guidelines also apply to masonry chimneys that are meant to serve a woodstove, a furnace or a boiler.

The Uniform Building Code ($40.75 from I.C.B.O., 5360 S. Workman Mill Rd., Whittier, Calif. 90601) contains basic requirements for chimney construction, and most states and municipalities have similar rules tailored to meet particular regional needs. I have a 1964 U.B.C. that reads the same as the 1984 Massachusetts State Building Code. It calls for a fireclay flue lining with carefully bedded, close-fitting joints that are left smooth on the inside. There should be at least a 1-in. airspace between the liners and the minimum

Bob Syvanen lives in Brewster, Mass. Photos by the author.

Illustrations: Christopher Clapp

Just below the roofline, the chimney wall shown above changes from 4-in. thick concrete block to brick. Block is much faster to lay up than brick, but it's usually used where appearance isn't important. The brick wall has been corbeled out against a temporary plywood form, enlarging the chimney above the roof and also centering it on the ridge. From this stage on, a good working platform on the roof is essential, right.

4-in. thick, solid-masonry chimney wall that surrounds them. Combustion gases can heat the liner to above 1,200°F, and the space between liner and chimney wall allows the liner to expand freely. Only at the top of the chimney cap is the gap between liner and wall bridged completely, and here expansion can be a problem. We'll talk about this later.

Chimney height is pretty much dictated by code. According to my Massachusetts Building Code, "All chimneys shall extend at least 3 ft. above the highest point where they pass through the roof of a building, and at least 2 ft. higher than any portion of a building within 10 ft." This rule applies at elevations below 2,000 ft. If your house is higher above sea level than this, see your local building inspector. The rule of thumb is to increase both height and flue size about 5% for each additional 1,000 ft. of elevation.

Exterior chimneys must be tied into the joists of every floor that's more than 6 ft. above grade. This is usually done with metal strapping cast into the mortar between masonry courses. No matter where your chimney is located, it should be built to be freestanding. The chimney can help support part of the wood structure, but the structure shouldn't help support the chimney. And there should be at least 4 in. of masonry between any combustible material and the flue liner.

The U.B.C. also has standards for flue sizing. For a chimney over a fireplace, the interior section of a rectangular flue tile should be no less than $\frac{1}{10}$ the area of the fireplace opening. If you're using round flue tile, $\frac{1}{12}$ the area of the fireplace opening will do because round flues perform slightly better.

Rectangular flue tile is dimensioned on a 4-in. module—8 in. by 8 in., 8 in. by 12 in., 8 in. by 16 in., 12 in. by 12 in., and so on. Standard tile length is 2 ft., though you can get different lengths from some masonry suppliers. When a mason talks about a 10-tile chimney, he usually means that it's at least 20 ft. high.

Using the flue-sizing formula isn't a guarantee that your chimney will draw properly. The chimney height and location, the local wind conditions, the firebox type, and how tight the house is are all factors that influence performance. John Hilley has found that on Cape Cod (at sea level), a 10-tile chimney atop a shallow firebox will work fine with flue sizing as low as 7% of the fireplace opening. Generally, short chimneys won't draw as well as taller ones. If you've got any doubts about what size flue tile to use, it's a good idea to ask an experienced mason or consult with your local building inspector.

Chimney construction—The fireplace chimney starts with the first flue tile on top of the smoke chamber of the fireplace (see the drawing on the facing page). As described in the article on pp. 94-98, the smoke chamber is formed by rolling the bricks of each course above the damper to form a strong, even, arched vault with an opening that matches the cross section of the chimney's flue tile. Since the smoke chamber will carry the weight of the chimney above it, it's got to be soundly constructed. If the rolled bricks of the chamber form an even, gradual arch, there should be no problems.

To begin the chimney, seat the first flue tile on top of the smoke-chamber opening in a good bed of mortar. Building paper or an empty cement bag on the smoke shelf will catch mortar droppings. You can remove the paper by reaching through the damper when you've finished the chimney.

Lay the chimney wall up around the flue tile, maintaining a 1-in. minimum airspace between tile and chimney wall. It's best to lay one flue tile at a time, building the chimney wall up to a level just below the top of each liner before mortaring the next one in place. Use a level to keep the liners and the chimney walls plumb. The mortar joint between tiles should bulge on the outside, but use your trowel to smooth it flush on the inside.

At every other liner, Hilley bridges the 1-in. airspace with two bricks or brick fragments that butt against the flue tile. These should extend from opposite sides of the chimney to opposite ends of the liner. This stabilizes the flue stack without limiting its ability to expand and contract with temperature changes.

If the chimney wall will be hidden behind a stud wall or in an attic, you can use 4-in. thick concrete block, which can be laid up much faster than brick. You can then switch to brick just below the roofline. Below the roof, chimney size is usually kept to a minimum to save space. If a larger chimney is desired above the roof, the chimney walls can be corbeled in the attic space (photo above left). The corbel angle shouldn't be more than 30° from the vertical (an overhang of 1 in. per course). A temporary plywood form set in place at the desired angle works well as a corbeling guide.

Coming through the roof—Roof sheathing should extend to within $\frac{1}{2}$ in. of the chimney, with structural members around it doubled and no closer than 2 in.

The next step in this part of chimney construction is to set up a good working platform on the roof (photo above right). Staging for

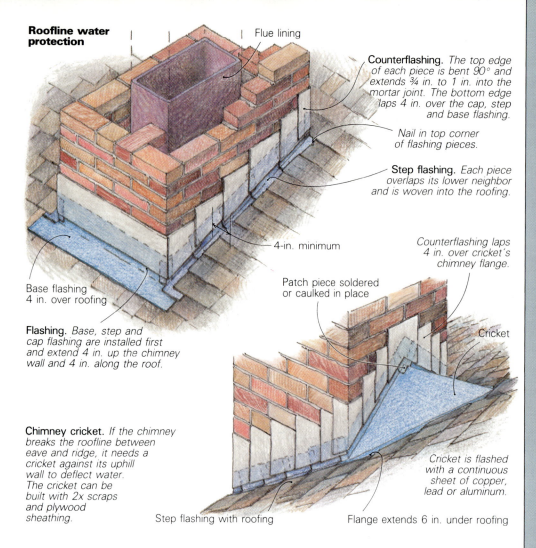

Roofline water protection

Flue lining

Counterflashing. *The top edge of each piece is bent 90° and extends ¾ in. to 1 in. into the mortar joint. The bottom edge laps 4 in. over the cap, step and base flashing.*

Nail in top corner of flashing pieces.

Step flashing. *Each piece overlaps its lower neighbor and is woven into the roofing.*

4-in. minimum

Base flashing 4 in. over roofing

Flashing. *Base, step and cap flashing are installed first and extend 4 in. up the chimney wall and 4 in. along the roof.*

Chimney cricket. *If the chimney breaks the roofline between eave and ridge, it needs a cricket against its uphill wall to deflect water. The cricket can be built with 2x scraps and plywood sheathing.*

Counterflashing laps 4 in. over cricket's chimney flange.

Patch piece soldered or caulked in place

Cricket

Cricket is flashed with a continuous sheet of copper, lead or aluminum.

Step flashing with roofing

Flange extends 6 in. under roofing

chimney work must be steady, strong and roomy. You can rent steel staging or build your own from 2x material. Either way, check the soundness of your staging well before you load it up with bricks and mortar. And before you bring any mud up on the roof, lay down a dropcloth to keep the roof clean.

To get to the staging, I put a ladder on the roof with its upper end supported by a ridge hook. It's a good idea to put some padding between the eave and the ladder rails so that the edge of the roof isn't damaged as you trudge up and down. You'll need a second ladder to get from ground to roof, unless the roof is steeply pitched. In that case, you can simply extend one ladder (if it's long enough) from ridge to ground.

Flashing—You're bound to find generously caulked flashing if you look closely at the chimneys in your neighborhood. I've lost track of the patch jobs I have done trying to stop chimney-flashing leaks. What sometimes happens is that rain gets blown in behind this flashing during a storm. Infrequent leakage doesn't mean the flashing job was poorly done, and caulking is a good stopgap in cases like this. But if your flashing leaks regularly in rainy weather, it probably wasn't installed correctly in the first place.

The Brick Institute of America (1750 Old Meadow Rd., McLean, Va. 22102) recommends flashing and counterflashing at the roofline. This creates two layers of protection,

with the counterflashing covering the base and step flashing on all sides of the chimney.

The Brick Institute recommends the use of copper flashing, but today you'll see more widespread use of aluminum, since it's much less expensive. Through-pan flashing (explained at right) is usually done with lead.

If your chimney straddles the ridge, first install the base flashing against the two chimney walls that run parallel with the ridge. Then step-flash the sides. Each flashing piece should extend at least 4 in. onto the roof, and is held with one nail through its upper corner.

Install counterflashing over the step and base flashing. The bottom edges of the counterflashing overlap the base and step flashing by 4 in. (drawing above left). The top edges of the counterflashing are turned into the masonry about ¾ in. They can be cast into the mortar joints as the chimney is built, or tucked into a slot cut with a masonry blade after the chimney is finished.

If the chimney is located against the side of the house or in the middle of a sloping roof, then you need to build a cricket against the uppermost chimney wall (drawing, above right). Otherwise, water will get trapped here, and you'll eventually have a leak. I use scrap 2xs and plywood to construct the slope, then cover the cricket with building paper and flash it with a large piece of aluminum or copper. The cricket flashing should extend 6 in. under the shingles and be bent up 4 in. onto the masonry, where it's covered by counter-

Through-pan flashing

Driving rainstorms can cause a lot of water to penetrate a chimney through cap and brick. Even when perfectly installed, conventional flashing can do little to stop this kind of penetration. The best way to drain out water that gets between the outer brick wall and the flue lining is to install through-pan flashing. Through-pan flashing is just what it sounds like—a continuous metal pan sloped from the flue lining to the roof. Weep holes between bricks just above the pan provide drainage.

There's some controversy about the effect that through-pan flashing has on the strength and stability of the chimney, since it breaks the mortar connection between bricks in adjacent courses. According to the Brick Institute, "if there is insufficient height of masonry above the pan flashing, wind loads may cause a structural failure of the chimney." This might be a valid warning, but I've never seen this kind of structural problem here on Cape Cod, which has its share of windy weather. And there's no arguing that through-pan flashing creates a more complete water barrier at the roofline than the conventional flashing scheme. Unless you live in earthquake or hurricane country, I can't see any reason not to use this system.

You can use copper or lead for through-pan flashing. John Hilley prefers lead because it's less expensive than copper, more malleable and generally easier to work. A utility knife will cut the stuff. Lead isn't supposed to last quite as long as copper, but I've seen 50 year-old lead pans that are still in good condition.

Building the curb—Before you can install the pan, you've got to build a curb where the chimney walls come through the roof. This is done with a combination of angled bricks or blocks and mortar, set in a form made from 2x lumber (top photo, facing page). The side walls of the curb should match the slope of the roof, and be around 1½ in. above the roofline. The lower and upper walls of the curb should be 4 in. to 6 in. above the roofline, parallel with it, and level. Like chimney walls, curb walls should be at least 4 in. deep.

Top the curb with a smooth layer of mortar, and the next day, rub the surface with a brick and round the corners to soften any sharp edges that could pierce the lead. Then install base, step, and cap flashing against the curb. With through-pan flashing, the pan takes the place of the counterflashing that is usually attached to the chimney walls.

The lead pan—If the joint between flue tiles falls just below or just above the roofline, then one piece of lead works well since you can roll it out on the base, find the outline of the flue, and cut holes for the next flue section to fit through. Alternatively, two or more sheets of lead can be used to make the pan. Just overlap the joints 6 in.

To determine the size of the pan, add 20 in. to the length of the chimney's lower wall and 24 in. to 32 in. to its side-wall measurement. Measure the side wall by following the angled curb with your tape measure. The chimney shown here is 64 in.

wide by 32 in. deep, and it required a sheet of lead 84 in. by 60 in. (7 ft. by 5 ft.). Lead sheets usually come in even foot widths and different lengths. For this job, Hilley trimmed a 6-ft. by 8-ft. sheet to size. These dimensions work for a chimney straddling the ridge. When the front and back walls of a chimney are on the same side of the ridge, the lead on the back or upper side of the chimney should be long enough to extend under two shingle courses plus one inch (more on a steep roof).

With the curb finished and flashed, and the lead cut to size, the next step is to install the pan. Roll the lead out over the curb and position it symmetrically (photo center left), then press down gently to find the outline of the flue tile.

Hilley cuts the hole for the tile about 2 in. smaller than the outside dimensions of the flue. Then he makes relief cuts in each corner and folds up the lead so that the flue can slide through it.

You have to be careful when handling the lead sheet; it's surprisingly easy to tear and puncture. The easiest way to carry a sheet is to roll it up. Never form lead with a hammer. Use your hand, a block of wood or a rubber-handled hammer handle. If you do pierce or tear the pan while installing it, pull the hole up above the surrounding pan so that water will drain away from it. You can also mend a hole by parging it over. As you position the pan, keep in mind that the object is to direct water away from the flue.

Once the flue tile that extends above the pan is in place, pull the lead up the sides of the flue to achieve the necessary outward slope. You can stiffen the top edge of the pan around the flue by folding it over. This helps to prevent sagging. After the lead is formed up tightly around the flues, parge the lead-to-flue joint with mortar.

Laying up the brick—Lay the first brick course over the pan in a thin bed of mortar (photo center right). Be careful to follow the curb under the lead. Make a few weep holes at the lowest points of the pan. You can do this by temporarily inserting a twig or rope in the mortar between bricks, or by leaving a vertical joint between bricks open. You'll have to cut the bricks that fit just above the sloping sides of the pan. Use either a mason's hammer, a chisel or a masonry blade. Once the courses on both sides of the ridge join, finish the chimney just as you would if there were no through-pan flashing.

Finish the through pan by creasing its exposed corners, trimming the pan edges and bending them down against the roofing. Before you start this part of the job, carefully lift the lead and sweep out any loose rubble. Creasing the corners so that they look nice is hard to do. The easiest way to start the crease is by slipping a short length of wood or angle iron under the pan at the corner. Hold the wood or metal edge so that it bisects the 90° corner, and gently start to form the lead over it with your hand. Then, if necessary, use a scrap 2x4 or a similar tool to form the lead into a more defined crease, as shown at right. Sharpening the crease should force the pan down against the roofing. A second crease, close to the first one, will force the pan down even farther. —B.S.

The curb for the through-pan, above, is a combination of bricks and formed concrete, built directly on the chimney walls where they intersect the roofline. Curb walls should be at least 4 in. thick. The side walls of the curb are sloped to match the roof pitch, and 1½ in. above the roofline. End walls are level, parallel with the ridge and 4 in. to 6 in. above the roofline. At left, a lead sheet is cut to fit over the flue tiles and carefully rolled and bent over the curb. This forms the through-pan that will drain water away from the center of the chimney and onto the roof. Below, the pan is parged to the flue, and the first course of the chimney cap is mortared to it, bearing squarely on the curb.

The last step is to crease the corners of the pan, trim its edges and bend them down against the roof. Wood blocks and gentle hand pressure are used in this final forming.

flashing. Once the cricket is finished and flashed, flash and counterflash the three remaining sides as mentioned earlier.

The cap—The chimney cap covers the airspace between flue tile and chimney wall, stabilizing the flue and directing water away from the rest of the masonry (drawing, below left). It's good to corbel out the chimney's top brick courses or to install a cap that overhangs the chimney walls. This will direct runoff onto the roof rather than onto the brickwork.

You can buy precast caps in a few sizes or cast your own in place. Installing or forming a cap is more complicated than it sounds because of the way the ceramic flue behaves. Cross-sectional expansion of heated flue tile is accommodated within the 1-in. airspace between tile and chimney wall. The concrete chimney cap bridges this gap, and if it's mortared directly to the tile, you're bound to have cracking and breaking problems. Even if the upper tiles stay cool and don't expand widthwise, the tiles near the fire will expand along their length, forcing the entire flue upward.

One way to accommodate vertical movement of the flue is to create an expansion joint between brick courses just below the cap. The topmost flue tile, cast to the cap, forces it upward as the flue stack expands. The cap's weight closes the expansion joint as the stack cools. Hilley creates his expansion joint by sprinkling a thin layer of sand on the brick course before laying down the mortar (photo top left). The sand prevents the mortar from adhering to the brick directly beneath it, but doesn't affect the bond to the bricks above.

The chimney top's first corbeled course creates an inside ledge that will support a form for the concrete cap (photo center left). You can use corrugated sheet metal or scrap wood for the form. If you use wood, as Hilley does, make sure it sits loosely on the brick ledge, and soak the wood in water before pouring the cap. This way, the wood won't swell and force bricks out of their bond.

The last flue tile should project above the level of the last brick course at least 2 in. so the concrete cap can slope down toward the edges of the chimney (photo bottom left).

Forming the concrete cap is the last step. Hilley casts the cap directly to the topmost flue tile, relying on the control joint several courses below to accommodate flue-stack expansion. An alternate method is to cast an expansion joint around the topmost flue tile. To do this, pack some ⅜-in. dia. backing rope around the top flue tile where the cap will fit, and then caulk the space with a flexible, non-oil-base sealant after the cap is cast. □

The chimney cap

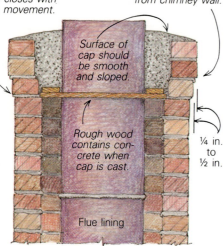

Expansion joint opens and closes with flue movement.

Top three to four brick courses are corbeled out to bring drip edge away from chimney wall.

Surface of cap should be smooth and sloped.

Rough wood contains concrete when cap is cast.

Flue lining

¼ in. to ½ in.

Capping the chimney. **Top, John Hilley creates an expansion joint with a loosely spread layer of sand beneath the mortar. Located several courses below the chimney top, this joint will widen and shrink with the normal expansion and contraction of the flue tile. The wood will be part of the concrete cap's bottom form. At center, the cap is cast between the flue and the top three brick courses. At left, the completed cap slopes away from the flue.**

The Point of Repointing

Renewing tired masonry joints can bring an old brick building back to life

by Richard T. Kreh

Brickwork deteriorates. This is inevitable but not, fortunately, irreversible. Replacing bad bricks and repointing with new mortar can restore a building's original integrity. And repointing, known as tuck pointing in the trade, is a job that can be handled by a careful homeowner. The resulting pride and satisfaction can be as substantial as the money saved.

In any repointing job, the first step is to determine exactly what the problem is. In some cases, the mortar simply has worn out. Old mortar often contains no portland cement, just lime and sand, with animal hair as a binder. These mortars can soften and break down over time, losing their ability to seal joints or adhere to brick. When this happens, they need to be replaced.

In other cases, improper flashing or structural damage to the roof has allowed water behind the brick. This water can be absorbed by the brick and mortar. As the wall dries out, the moisture escaping can cause the bricks to chip and the mortar to pop out. This problem is even more severe in colder climates where the expansion and contraction of freezing and thawing water accelerates the process. Sometimes the ground beneath the wall has settled. This can cause cracks in the wall, which might eventually require rebuilding. Before any actual repointing

---

Richard T. Kreh has over 33 years of experience as a mason and author of masonry books.

is done, find out why it needs doing, and correct these conditions first. If you don't, all the effort and expense of repointing will be wasted.

Preparation—Spring and fall are usually ideal times to do repointing work, because the best temperature for curing mortar is 70°F. Repointing can't always be done when conditions are perfect, but if you have a choice, arrange to do the work when they are at their best.

Never repoint when temperatures are likely to dip much below 40°F unless you can protect and heat the wall. Mortar that freezes before it cures will be brittle, and it will pop out when the wall warms up. Don't fall into the trap of repointing on a warm winter day and leaving the mortar to freeze overnight. If you hire a contractor to do the job, remember that he will schedule it to suit himself. He has a living to make year-round. Be sure to agree in writing that he will not attempt to work in freezing temperatures.

In hot weather, try to work on the shady side of the building to keep the mortar from drying out too fast. You might need to keep the wall damp (not soaked) with a fine mist from a garden hose or tank sprayer. (Never wet a wall in cold weather, because a freeze would cause the mortar to pop.)

For years, sandblasting has been used to remove dirt and old paint before repointing. Sandblasting is the continuous bombardment of

a masonry surface with abrasive particles sprayed through a nozzle in a high-velocity stream of air. Silica sand, the aggregate often used, cuts away part of the brick's surface. Cutting through this outer surface can cause permanent damage. Most disasters occur when inexperienced operators do their own sandblasting. If your brick face needs to be cleaned and you decide to use this method, hire a reputable contractor who specializes in such work.

Clear, liquid silicones can be applied to brickwork after it has been sandblasted and repointed. When reapplied periodically, they will protect the brick face for years. This treatment is especially recommended for historic buildings. Silicones are available from building-supply dealers. Follow manufacturers' instructions when applying them.

Repointing—You can't repoint without making a mess. There will be dust, dirt and mortar droppings to contend with. If you plan ahead, though, you can make cleanup a lot easier. Construction-grade rolls of 4-mil polyethylene plastic are ideal for protecting windows, doors, and flowerbeds and shrubbery near the house. Work on one area of the house at a time, and clean up as you go along. If you have to rent scaffolding, do all the high work first, so you can save money by returning it as quickly as possible.

The traditional method of removing mortar

Repointing Mortar

You don't need an especially strong mortar for repointing work. Repointing mortars need to bond with the old brick, seal the joints and match the original mortar as closely as possible in both strength and appearance. This is best done by a mortar with a much greater percentage of lime than of portland cement. Mortars with a high percentage of portland cement are often harder than brick; the brick gives under stress rather than the mortar. The result is *spalling*, a cracking and flaking of the brick face. High-lime mortars form a resilient cushion on which the brick can rest. They shrink very little, hold water well during repointing, and are easily worked. They can also heal small hairline cracks when the wall is moistened. During this process of autogenous healing the hydrated lime dissolves in water and is then recarbonated by the atmosphere's carbon dioxide, sealing the crack.

For an excellent all-around mortar with good bonding ability, use Type S hydrated lime, Type 1 portland cement, washed bank sand and potable water. Mix 1 part Type 1 portland cement to 2 parts Type S hydrated lime to 8 parts washed building sand. If the sand is coarse (sharp), add an extra half-part lime to improve workability and bonding. This mortar will test out at about 300 psi after it has cured—strong enough. —*R.K.*

The author repoints a brick wall.

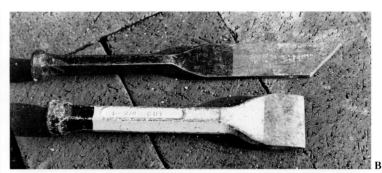

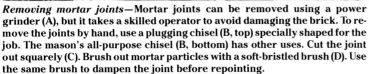

Removing mortar joints—Mortar joints can be removed using a power grinder (A), but it takes a skilled operator to avoid damaging the brick. To remove the joints by hand, use a plugging chisel (B, top) specially shaped for the job. The mason's all-purpose chisel (B, bottom) has other uses. Cut the joint out squarely (C). Brush out mortar particles with a soft-bristled brush (D). Use the same brush to dampen the joint before repointing.

joints is to use a hammer and chisel. Today, power grinders are often used to speed up the process (photo A). Even with a trained operator using the machine, however, it is all too easy to damage brick edges. The old way is best; repointing is not a job that can be rushed.

To remove mortar by hand, use a joint (or plugging) chisel (at the top of photo B). It has a specially-tapered blade that cleans mortar from the joint without binding or chipping the brick surface. It should be available at any store that stocks masonry supplies. Cut the joint out to a depth of ¾ in. to 1 in., depending on the depth of the deteriorated mortar (photo C). Cut it out square, not in a V shape.

After cutting, measure with a rule to be sure you've reached the correct depth. Brush all dust and mortar particles out of the joint with a whitewash or wallpaper paste brush (photo D). Next, use a wet brush to dampen the joint. Don't soak it, or the new mortar will not adhere. The idea is to prevent the repointing mortar from drying out too quickly.

Tuck-pointing mortars should be prehydrated to improve workability and reduce shrinkage. To do this, mix the cement, sand and water thoroughly in the proper proportions. Then mix again, adding enough water to produce a damp (but unworkable) mix that will retain its form when pressed into a ball in your hand (photo E). Leave the mortar in this dampened condition for an hour or two, then add enough water to bring it to a consistency that is somewhat dryer than regular masonry mortar, as shown in photo F.

When the mortar is ready, apply it to the joint. Use a thin, flat piece of steel, called a striking tool, to press the mortar firmly into the moistened joint, filling it out (photo G). If the joint has been cut out to a depth greater than 1 in., apply new mortar in two layers. Allow time for the first layer to set before beginning the second.

Don't tool a finish on the joint until the mortar has set enough so that it won't smear. If your thumb leaves a clear impression on the mortar, it's time to tool. Doing so too early will cause the joint to streak or sag. Waiting too long will result in a black mark, which is caused by a reaction between the slicker and the chemicals in the mortar. (This is known as "burning the joints.")

Finish the joint by tooling a flat, concave or grapevine shape on the surface. (A grapevine is formed with a steel tool that leaves an indented line in the mortar.) With a little practice, it won't take you long to get the hang of finishing. After the mortar has set well enough to prevent smearing, brush off the wall to remove all mortar particles (photo H).

Replacing bricks—In most repointing jobs some deteriorated brick will have to be cut out and replaced. Use a plugging chisel to remove the mortar joint, and an all-purpose mason's chisel to cut out the soft brick (photo I). Brush away all dust particles. Use mortar to butter the replacement brick and all sides of the wall cavity in which it will be placed. This prevents any voids in joints after the new brick is inserted. Push the new brick in place and point the mortar joints (photos J-L).

You should make an effort to match the original brick. Look around the site or in adjacent outbuildings for bricks left from the original construction. If they haven't softened with age, they would be the best choice. You can also try your local building-supply dealer. Many companies make new bricks that look like old, handmade ones. Take a sample of the original brick with you and compare for size as well as color.

Salvaged brick may also work, but be careful. If reclaimed bricks are soft, they won't last long once exposed to the elements. You can make sure a brick is hard enough to use in repointing by holding it in your hand and tapping it lightly with a hammer. A soft brick will give off a dull thudding sound; a hard one, a metallic ring.

Cleanup—If you've been neat, only a little cleaning will be necessary after repointing. Wait at least two weeks to let the mortar cure completely, then scrape particles and clumps off the wall with a wooden paddle or a chisel. Hose down the wall and give it a good scrub with a stiff-bristled brush.

If stubborn stains remain, or you have splashed mortar down the wall, you may need to scrub it with a solution of 1 part muriatic acid to 20 parts water. Wear hand and eye protection, cover the flowers and shrubs near the house, and hose off the wall when you're done. □

Mixing mortar and repointing—Basic tuck-pointing mortar is composed of portland cement, lime and sand. Mix the dry ingredients in the proper proportions, then dampen until a ball of the mix squeezed in your hand retains its shape (E). This prehydration improves the mortar's workability and reduces its shrinkage. After an hour or two of prehydration, add more water to bring the mix to its final consistency, somewhat dryer than regular masonry mortar. After wetting the joint, press the tuck-pointing mortar firmly into it with a slicker (G), then tool the finish. Once the mortar has set and won't smear, brush off any particles (H).

E

F

G

H

I

J

K

L

Replacing deteriorated brick—Use the plugging chisel to remove the mortar around the bad brick, then the all-purpose chisel to cut out the brick itself (I). Brush away all mortar and brick particles (J). Give both the replacement brick and those still in the wall a thick coating of mortar, and press the brick into place (K). Then point the mortar joints (L).

Repointing Old Brick

Using the wrong mortar can do more harm than good

by Jean Dunbar

When I left at noon, the mason was getting ready to repair some failed mortar on my hundred-year-old brick house. By the time I came back in the afternoon, he had finished patching some of the joints with portland-cement mortar—and set up my house for permanent and irreversible damage. When used on brickwork built before World War II, the improper mortar can do the job of a wrecking ball—more slowly, but just as well.

Replacing failed mortar with sound mortar is called repointing. Repointing is more than mixing up a batch of mortar and smearing it into the cracks. The entire process includes removal of the failed mortar and, often, careful cleaning. But the key to successful repointing is using the right mortar. Old mortar is different from modern mortar because old brick is different from modern brick. A look at how old mortar works with old brick explains why portland-cement mortar can destroy an old building's bricks.

Before the 1870s, when brick-cutting steel forms came into use, bricks were made by hand in individual molds. These hand-molded bricks are less uniform than the steel-cut ones. Older bricks are also softer than their modern counterparts. Wood-fired kilns, used before this century, fired bricks unevenly, producing bricks of variable hardnesses. The outer $\frac{1}{8}$ in. of brick, which is exposed to the highest temperatures in the kiln and later to gradual hardening by air, is the brick's hardest part. This shell, known by masons as the patina, protects the softer inside of the brick from the ravages of water and weather.

Pre-20th-century mortar was made of river sand and burned lime. Like the hand-molded bricks, lime mortar was soft, with a low compressive strength measured in the hundreds of pounds per square inch. Modern mortar, on the other hand, is made with portland cement and is designed to withstand pressure from thousands of pounds per square inch. It is hard and rigid.

During temperature changes, when a wall expands or contracts with heat or cold, the weakest, softest areas of the wall surface must absorb the change. Because lime mortar is softer than brick, it acts as a sacrificial element, absorbing the bricks' expansion and contraction; the mortar will erode, crack and fall away long before the bricks. When hard portland-cement mortar is used to repoint old brick, the brick is softer than the rigid mortar, so it—not the mortar—gives way. Spalling (the breaking off of the outer surface of

Part of the process. **If a building's bricks are badly soiled, it's best to clean them gently with water. Never sandblast old bricks because sandblasting strips off the protective surface of the bricks.**

the brick) and cracking of the bricks result (top photo, facing page). The use of hard mortar is the most common error in the repointing of old brick buildings.

John Friedrichs is a mason in Lexington, Virginia, whose forte is restoring historic brickwork. Friedrichs' specialized methods require some understanding of old brick and lime mortar, but they require few specialized tools and are appropriate for any brick building. Using Friedrichs' common-sense techniques, anyone who owns a masonry hammer, a tuck-point chisel, a pointing tool and a trowel (photo, p. 116) can repoint an old brick building without damaging its bricks.

Cleaning can be the first step—Any brick-cleaning method must leave the brick's hard, protective surface intact. Sandblasting was long considered to be the best way to clean old brick, but sandblasting strips off some of the brick's protective surface, exposing the soft core of the brick to the weather. The mottled surface that results permanently spoils the building's appearance. Worse, with its protection gone, the brick becomes porous. Porous brick absorbs water, and when this water freezes, the brick spalls.

Highly pressurized water can have the effect of sandblasting. It can literally carve the brick. Strong chemicals can damage the patina on contact. If not washed off completely, they can damage the brick over time. Consequently, removal of graffiti or paint—which usually requires acidic cleaners and/or the use of poultices designed to hold solvents in place—should be undertaken by a professional restorer (see *FHB* #84, pp. 72-75).

To clean simple soil stains, Friedrichs first tries to determine how the brick got dirty in the first place. Brick is most often soiled as a result of water splashing back onto it or from airborne soot on buildings that were once heated by coal.

Friedrichs removes these types of soil by simple rinsing. The work must be done when there is no danger of frost. Using a power washer adjusted to low pressure, he directs a light mist of water at the top of the wall. As the water runs down, it gently bathes the brick. For final rinsing, Friedrichs reduces the pressure further so that the water gently floats away any remaining loosened soil. Gentle scrubbing with a soft scrub brush and clean water can substitute for the power washer, and cleaned areas can be rinsed with a garden hose (photo facing page).

Analyzing the original mortar—Friedrichs tries to match the old lime mortar as closely as possible. He first conducts an on-site test that determines the particle size, the color and the silt in the building's original mortar. For his test, Friedrichs removes half a cup of clean mortar from the building. He then pulverizes the mortar. Working outside, wearing goggles and chemical-resistant gloves, Friedrichs checks to see which way the wind is blowing so that he avoids chemical fumes. He then puts the pulverized mortar into a quart jar and covers the mortar with a 3% solution of muriatic acid.

With the top of the jar off, Friedrichs allows the mixture to bubble up. When the bubbles subside, he sloshes the mixture around, adds more solution and allows the contents of the jar to bubble up again. When the bubbles subside, he adds at least one cup of water to stop the reaction. Dirt now floats in the water while sand remains on the bottom of the jar.

Friedrichs saves both residues. He pours the water/dirt solution through a paper coffee filter. He saves (in the filter) all the particles that were suspended in the solution and leaves the sand, with as little water as possible, at the bottom of the jar. The sand will be matched. The dirt helps gauge how dirty the original sand was so that the color and the texture of the new sand can be adjusted accordingly.

Developing a sand sample—A mortar's color is mostly determined by the color of the sand used in the mortar. So you can do a credible job of matching mortar by using whatever commercially available sand best matches the sample from the original mortar. Friedrichs has a library of different-colored natural or commercial sands that are available. But for an accurate restoration, he begins matching old mortar by finding the source of the sand that was used to mix the mortar when the building was constructed.

Before the age of dump trucks, most building materials came from within a few miles of the building site. The older the structure, the closer

Spalling. **Portland-cement mortar, when used to repoint old bricks, can do more harm than good. The hard mortar doesn't expand and contract during seasonal temperature changes and can cause the outer surface of the soft bricks to break off—a condition known as spalling.**

Removing failed mortar. **Mason John Friedrichs uses an air tool to remove only the center of each failed mortar joint in this wall. All the mortar that adheres to the bricks themselves is carefully cleaned out by hand.**

The blob. **You'll know that the proper proportions of sand, lime and water have been added to the batch of mortar when a blob of mortar will not fall off a trowel when the trowel is held upside down.**

Tools of the trade. From the top: A mason's hammer; a tuck-point chisel; a pneumatic air chisel and three different size carbide bits; an air nozzle for blowing out cleaned mortar joints; a trowel; and on the bottom, a pointing slicker.

the original masonry materials were to the site. Friedrichs looks for the river or stream closest to the building site. He examines the outer shores of the stream's bends, where, because the current is the slowest, sand beaches tend to develop.

Friedrichs takes a sample from the sand he finds. He puts a half cup of this sand in a clear jar with water. After shaking the jar, he allows the sand to settle for half an hour. To test the dirt, just as he did when he reduced the building's original mortar to its components, Friedrichs uses a filter to separate the sand and the dirt. He then allows both pairs of samples—the original mortar and the river sand—to dry. (The hood of a truck is a convenient, warm place for drying them.)

Using a good-quality 4-in. magnifying glass, he examines the two pairs of samples, comparing the particle size, the color and the dirt to be sure they match. If the sample is lighter than the original, it is probably too clean: It lacks dirt that was present in the original sand. Very gradually, in measured amounts, Friedrichs adds either pure clay from around the building (sifted through a piece of screen to remove debris) or orange masonry sand until the color matches the original color of the sand in the building.

If an exact match proves impossible, a mortar that is slightly too dark will be less obvious than one that is slightly too light. Friedrichs records the substances added and the amounts in which

they were added so that the correct proportions can be maintained in larger mortar batches.

Mixing the mortar—The matching sand must be screened to remove debris before it is mixed into the mortar. Friedrichs uses a 2-ft by 3-ft. screen, with 2x4 frame and ¼-in. rat wire. With a 2x4 he props up the screen to an angle of about 60°. The steeper the slope of the screen, the finer the screened sand will be. Because old mortar was coarse, the slope must be gentle. Once the screen is set up, simply throwing shovels of sand against the screen removes undesirable material.

Screened sand and lime, together, make mortar. Friedrichs starts mixing mortar by pouring 3 qt. of dry sand and 1 qt. of dry lime into a wheelbarrow. (Too much sand creates a sharp mix that grates when it is handled instead of sliding smoothly). Lime is caustic. When working with lime, always work upwind when pouring, wear safety glasses and never touch your eyes when handling it because lime can blind.

After Friedrichs folds the lime and sand mixture together with a hoe, he adds water to the two ingredients—a little at a time. The mortar has reached the correct consistency when a blob of mortar held upside down on a trowel adheres to the trowel (bottom right photo, p. 115). It's better to err on the dry side when mixing mortar: It can always be dampened with water.

When the mortar is the correct consistency, Friedrichs verifies that the new mortar will match the old. He scrapes off the weathered patina from a small area of mortar that doesn't need repointing. Adjacent to it, he removes some mortar from a joint (see description below) and fills it with the sample mortar. He lets the sample sit until it is dry to the touch and resistant to fingernail scratching. (A hair dryer, a heat gun or a heat lamp will speed the process.) Because the mortar color on all sides of the building may not be exactly the same, the test is repeated on all sides of the building before repointing begins.

Removing failed mortar—Before cleaning out any damaged joints, Friedrichs first decides which joints really need repointing. A sound, if slightly worn, mortar-struck joint is smooth. With real wear, however, the surface—the hard lime skin—of the joint wears off, and an eroded, granular look appears.

This erosion can best be seen from below. When the bottom edge of the bricks above a joint are visible from below, the mortar has eroded, and the joint requires repointing. Repointing is particularly needed at the corners of the building, where erosion is accelerated by wind, rain and dripping water.

To preserve as much of the original work as possible and to minimize cost, Friedrichs always

repoints as few joints as possible. Where cracks have been produced by settling, he cleans out only a distance of half a brick on both sides of the crack. (Repointing will not cure structural problems, which require structural solutions that are beyond the scope of this article.)

Many masons use a grinder to remove the mortar from joints that will be repaired. Unfortunately, grinders can tear into soft, old brick, causing irreparable damage. They also cut along a straight line and create uniform-looking joints that destroy the appearance and the character of irregular, old brick.

Instead of a grinder, Friedrichs uses air tools—miniature hand-held jackhammers designed for use by sculptors (photo facing page). Even though an air tool is at least four times as fast as removing mortar with a tuck-point chisel and a hammer, an air tool preserves the irregularity of the brick because different-size carbide bits can be used to accommodate variations in the size of the joint.

Friedrichs uses his air tool to remove only the center of the mortar in the joint (bottom left photo, p. 115). At least 10% of the mortar—the mortar that adheres to the bricks themselves—is cleaned out carefully by hand raking with a pointing slicker (photo facing page).

Friedrichs uses air tools and air-tool supplies from Trow and Holden (P. O. Box 475, Barre, Vt. 05641; 802-476-7221), the company whose air tools were used to carve the Lincoln Memorial. They are reasonably priced for anyone who has some volume of masonry repair to undertake: $1,200 to $1,300 buys a complete setup, which includes an air tool, a couple of bits and an air compressor. Air tools also require safety equipment. Friedrichs wears a hard hat, a safety line, safety glasses and a respirator—and he insists that his employees wear them.

Friedrichs removes as much bad mortar as possible with the air tool. He removes mortar to a minimum depth of ¾ in. Sound mortar that doesn't need to be removed will resist the action of the air tool.

After all the joints in an area have been roughly cleaned, he dampens them with water from a pump sprayer to soften the remaining mortar. After dampening, each joint must be detailed by hand. The sound mortar remaining in the back of the joint must be squared off with a pointing slicker to form a flat surface that is perpendicular to the brick. Friedrichs also scrapes the bricks with the blade side of the pointing slicker to remove the thin layer of mortar on the surface of the brick. When the back of the joint has been squared off, Friedrichs blows the debris out of the joints with compressed air.

Once all detailing in the area under repair is completed, but before repointing, Friedrichs again dampens the entire area with the pump sprayer. Old brick is very dry. If it is repointed when dry, it acts as a sponge, drawing the water out of the wet mortar. If the brick wicks away the water, the mortar will dry too quickly and unevenly to cure properly, and it will be weak. For the same reason, when Friedrichs re-lays old bricks, he submerges the bricks in a tub of water for at least a minute or two before laying them.

Pointing, striking, brushing. **After cutting a sliver of mortar from the blob on his trowel with a pointing slicker, Friedrichs first fills the head joints (verticals) and then the bed joints (horizontals) in a section of wall he has prepared for repointing (top photo). In the bottom left photo, he angles his striking tool to cut off the excess mortar and to mimic the angle of the original joints in the building. The two top-right head joints and the top-right bed joint have already been struck. In the bottom right photo, Friedrichs uses a brush to remove the mortar crumbs left after striking.**

Repointing—One of the beauties of lime mortar is that it dries slowly. It remains usable for four to five hours with only minimal retempering with water. And if the mortar is kept in the shade, it can be mixed in a quantity large enough for about a half day's work.

After mixing up a batch, Friedrichs scoops some mortar onto a trowel and cuts a modest sliver of mortar with a pointing slicker, cutting toward the outer edge of the blob of mortar. He wipes one sliver of mortar at a time into the joint until the joint is full (top photo, above). Neatness is efficient, but it is all right for mortar to overhang the joint. A carefully prepared mortar mix is relatively dry and leaves no residue. At a minimum, each joint must be filled flush to the face of the bricks. To ensure a clean finish, Friedrichs first fills the vertical (head) joints and then the horizontal (bed) joints.

The final finish of the repointed joints must match the original finish. Before the mortar dries completely, the surface of each joint must be undercut slightly with the edge of the pointing slicker. This is called striking the joints. Instead of simply smoothing each joint with the flat of the tool, Friedrichs angles his pointing slicker to mimic the original work (bottom left photo, above).

In the original work on any old building, all head joints on a given wall are angled in the same direction: Either the left side of the mortar joint is flush with the left-hand brick (while the right side of the joint is slightly undercut), or the right side is flush with the left-side undercut. Bed joints are treated the same way. Either the mortar is flush with the upper brick while the bottom edge of the joint is slightly undercut, or the opposite is true.

Repointed mortar must follow the original strike pattern. Otherwise, the sun hitting a wall produces a confusion of shadowlines. Holding a pointing tool with its tip at the top of what will be the shadowline, Friedrichs cuts back the mortar. Pressing firmly and evenly, with the tool slightly angled, he strikes each joint.

When all of the repaired joints in the area have been struck, Friedrichs uses a relatively soft, long-bristled mason's brush (bottom right photo, above) to remove mortar crumbs.

Successful restoration, properly executed, stabilizes the skin of a historic masonry building, protecting it against the elements and reversing damage caused by weather, poor maintenance or earlier ill-conceived repairs. Thoughtfully executed work restores the appearance of the building so subtly that the building somehow looks better—without anyone, except the restorer, knowing just how and why.   □

Jean Dunbar is a freelance writer in Lexington, Va. Photos by the author.

Matching Existing Brickwork

Perseverance solves a remodeling mystery

by Steven Spratt

It started off small and simple, the way a lot of remodeling jobs do. Our neighbors, impressed by work we had done on our own house, asked us to update the kitchen of their 40-year-old home. They had raised twelve children in it, and the kitchen showed it. Furthermore, to accommodate the holiday get-togethers of a dozen adult children plus all *their* kids, a bigger kitchen was necessary. Once the decision was made to add more space, an architect and engineer were brought on board.

A mystery unfolds—Throughout the planning of the project, the owners had made it clear that no matter what was done to the house, the original architectural style must be preserved. This posed an interesting problem because the entire home was constructed of some unknown kind of masonry (bottom photo, facing page). After contacting many masonry contractors and all the local suppliers, it became clear that matching the material was going to be very difficult. Pinkish-tan in color, the bricks were more uniform in shape than traditional adobe bricks. Sandy and pockmarked with tiny voids, they were some sort of unfired cement-based product, with color integral to the material used to make them. Vertical striations on the sides and face of the bricks further clouded their identity. And as both the structure of the building and its exterior finish, they were hollow inside to accept concrete filler and steel rebar.

During the demolition phase of the project, we tried to salvage what bricks we could. This proved impossible. Attempts to split the bricks apart only shattered the soft outer brick, leaving the hard concrete core intact. Bricks this soft (even if we could have found them) would not be acceptable structural materials anyway, because the addition had to meet stricter seismic codes than did the original home. We finally decided to frame the addi-

Days of experimentation finally led to a mix that would, after drying, match the existing brickwork. The wet mix was shoveled into forms, packed through and then left to dry in the sun. In the photo above, sheet metal has been press-fit into the molds to form the rounded corners for sill bricks.

tion with wood and use some sort of masonry veneer as the exterior finish. That's when the fun really began.

The widow remembers—In previous conversations with the current owners, we found out that the original owners were still in the area. When we called to query them about the bricks, we learned that the husband had passed away some years ago, but his widow vaguely remembered that the bricks had been hauled in from Fresno. This was our best lead so far—maybe we'd find some obscure factory making bricks there.

Dozens of phone calls taught us that Fresno had a very large and active sand, gravel and ce-

ment industry, with many small companies making all kinds of products, but nobody made anything like the bricks we wanted. To say we were discouraged was an understatement.

That weekend, Fresno was all I could think of. Had the factory that made our bricks disappeared? Or was the old lady just mistaken about the source of the originals? By Saturday night I could stand it no longer—I asked my wife if she'd like to go on a highly speculative brick hunt with me.

At last, a smoking slab—Fresno was a four-hour drive away, and when we got there it was already hot and dusty. We made our way to a couple of materials yards, but most were closed for the weekend, and the one that was open had nothing to offer. It was getting late in the afternoon, and we were tired. As we pulled into a gas station in a town called Friant Dam to gas up for the trip home, I was absolutely astonished to see an entire motel constructed of the bricks I was looking for. The material was exactly the same, right down to the trim pieces used under the windows.

I could hardly contain my excitement. For months I'd searched for anything that would remotely resemble our bricks, and now I was standing in front of an entire building made of the stuff. A quick look around, and I spotted four other buildings made of it; we had to be close to the source now. After a few inquiries we discovered that a small brickmaking company had once existed in the area, but had burned to the ground and no one knew exactly where it had been. We were told to check up the road at a quarry operating there.

So up the road we went. We came to a place called Four Corners: a large shed, a small store, an outhouse, two dilapidated houses and some ruins of about three more houses. All but the tin shed were made of "the brick." At the store I was told the quarry was closed, but that the watchman lived in the trailer beside the shed. Walking around the back of the store I stumbled over a pile of debris in the high weeds. As I looked closer, cursing my clumsiness, my eyes bulged: I was staring at a neatly stacked pile of perhaps 50 perfectly shaped, perfectly colored (if slightly weather-worn) window sill bricks of exactly the type I needed.

This had been a brick factory alright—the burned-out building slab was still there, as was some rusting and unidentifiable machinery, but except for the sill bricks there was nothing I could use. I walked on up to the watchman's trailer and knocked. He was watching the Forty-niners on TV and wasn't too happy to see me, but after hearing my story he began to warm up.

The plot thickens—"Friantite," said the watchman. "That's what they call this stuff here," and he gave the dirt a kick. "It's a pumice." He told us that the reddish color is from iron in the upper strata of the quarry. Deeper within the quarry the color fades—it goes from red to tan to grey to white. The white stuff at the deepest part is slick—like talcum powder—and called pozziline. Concrete companies use it as a lubricant and strengthener. It's also used as a vehicle for certain pesticides. That was all the fellow could offer, but he gave us the phone number of his boss.

It was late, so we called it a day and headed home. What had I expected find, anyway? A material yard stacked fence high with everything I needed? No, I was lucky to find anything at all, and by the time we reached home that night, I knew what we had to do next. We were going to make our own bricks.

A pit and forty of the best—Monday morning I called the watchman's boss, Mr. Richart, who owned the quarry. He knew nothing about the brick factory, but did suggest that we might find enough raw material on his property. If so, we were welcome to it. I promised to return the following Saturday.

Tuesday I met with Keith Armstrong, our architect, and explained my plan to make the bricks. We would drive to Fresno in a large truck, find a suitable deposit of Friantite, load the truck and bring it back to the job site. Keith jumped at the chance to accompany me.

We left early on Saturday with picks and shovels in the back of the truck. We met Mr. Richart around noon, and he rode with us in the front seat of the dump truck as we headed for the quarry.

It was another hot afternoon. The quarry was probably a mile square and 100 feet deep, and by the time we walked around it we were covered with dust and sweat. The walls of the pit were hard as rock, but once broken off and crushed, the material powdered up like dust. In the roadways where trucks had driven, stepping in the dust sent it rippling like waves in water.

After a thorough search we found a spot where the color was nearly what we needed. This is where we decided to start. Upon see-

Evidence of success. The walls of the addition (at right in the photo above) are difficult to distinguish from those of the original house.

Faced with an addition to build and an existing house built with bricks of unknown origin (photo above), there was no choice but to make new bricks. That was the easy part; the hard part was finding the raw materials.

ing our picks and shovels, Mr. Richart laughed. "You'll be a while in the hot sun if you plan to dig that rock by hand," he said. "I have a better way." With that, he left. In a few minutes the roar and dust of a huge loader came toward us. In one scoop, the loader pried loose enough material to eliminate hours of backbreaking work. The stuff came out of the hill in big rocky chunks. As I was wondering how we would grind it into usable material, Mr. Richart surprised us once again. After pulling up alongside our truck and dumping the whole load out on the hard ground, he proceeded to run it over with the loader. After half a dozen passes he peered from the cab and said, "Is that fine enough?" It sure was.

After loading up, we threw a tarp on top of our precious load of material. And on top of the tarp we threw forty of the best sill blocks we could salvage from the pile I had found.

Bullion and plastic explosives—Now we had the raw material for making the bricks, but we still had to experiment with the manufacturing process. The recipe that finally proved to meet all our criteria was, mixed in order: water—2 parts to start; concrete colorant—as

needed; sand—4 parts; friantite pumice—4 parts; crushed aggregate—2 parts; and #1 white portland cement—2 parts. When these ingredients were mixed well, we added more water (between one and two more parts) until the mix passed a ⅛-in. slump test (meaning that the bricks slumped down in height no more than that when we lifted the forms). We used a hand cement mixer and kept it rolling for a minimum of ten minutes per batch—or, as our man Fred Pokorney put it, "You can't mix it too long."

Although the Friantite itself was very close in color to the old bricks, the mix changed color when the other ingredients necessary for strength and texture were added. Eventually we solved the problem by a process of trial and error. We made a large series of samples, labeling each with the precise type and amount of colorant used, before we hit upon just the right balance of colorant. New bricks dried to their final color a full week after being made.

The first forms we tried, fabricated from sheet metal, were intended to work like muffin tins—but they didn't. Even with plenty of refined diesel fuel used as lubrication, the bricks would not come out. The solution Fred came up with was much simpler: 2x4 frames divided to make five bricks at a time (all that one man could easily handle). For the curved corners that we needed on the trim pieces, we added pieces of sheet metal to the forms (photo, facing page).

As we started production, we were careful to measure the ingredients precisely to keep batch variation to a minimum. As for matching the texture of the original bricks, we got the necessary bubbles and striations by refraining from too much tamping or cleaning of the forms. The freshly made bricks were covered with plastic for the first 24 hours to keep the surface moist for even curing. After a day, they could be stacked, and after a week they were difficult to break with a hammer. We needed only 600 bricks, but we decided to make plenty of extra once we got going—just to be sure. Production averaged about 75 bricks per day, and soon we had a driveway full of bricks curing in the late autumn sun.

After spending this much time and effort getting the bricks, we wanted a mason who would treat these babies like gold bullion wrapped with plastic explosives. As it turned out, the owners knew a mason of good reputation, and we watched with increasing trust as he laid them up. The match with the old bricks was excellent.

The exterior walls of the kitchen addition now blend quite unremarkably with the older portions of the home, giving no hint of the labor behind them (top photo). Was it worth all the effort? Who knows—but the owners got what they wanted, and we have the satisfaction of having solved a seemingly unsolvable problem. □

Steven Spratt is president of Spratt and Co. Builders, a general contractor specializing in custom construction and restoration.

Chimneys

Venting a fireplace or furnace is the first purpose of a chimney, yet some chimneys rise above this humble, mechanical role. Embellished with plasterwork and stepped brickwork, they stand against the sky, monuments to the inventiveness of their makers. Left and bottom: The Nunan Mansion in Jacksonville, Ore., was architect designed for an Irish immigrant. The entire house was precut in Tennessee,

Santa Barbara, Calif.

Salem, Mass.

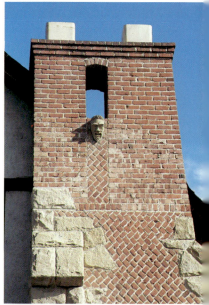

Santa Barbara, Calif.

Photos above and left: R. E. Grossman

shipped to Oregon in 14 boxcars and assembled in 1892 for $7,792. This page, right and bottom: The Kerr house in Grand Rapids, Ohio, was built in 1885. An example of early Queen Anne style, it features stained glass and butternut woodwork. It was recently renovated for use as a health retreat.

Above and top, Cape May, N. J.

Photos above and right: © Andrew Gulliford

Fireplaces

This see-through fireplace (facing page) over-looks both a bedroom and a bath. Designed and photographed by Obie Bowman. Built by Duncan Masonry.

A cast-concrete fireplace (photo above) supports a roof. Designed by Will Bruder. Built by Dave Platt. Photo by Paul Spring.

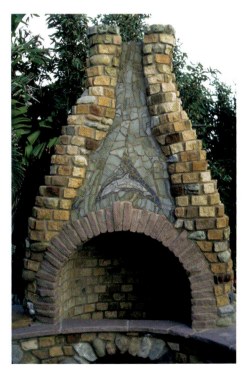

Conventional bricks (photo above) give way to clinkers and slag. Designed and photographed by Obie Bowman. Built by Duncan Masonry.

A mosaic porpoise leaps over the firebox of an outdoor fireplace (photo left). Designed, built and photographed by John August.

Black granite and galvanized-steel stripes border a room-divider fireplace (photo above). Designed and photographed by Obie Bowman. Built by Redhorse Constructors.

Zero-clearance fireplaces (photos right) are customized with pine trim and simple drywall detailing. Designed by Dennis Wedlick. Built by Jon Lynn Jones. Photo by Michael Fredericks.

Tightly fitted sandstone veneer covers a fireplace enclosure (photo left). Designed and built by Leo Blickley. Photo by Scott Gibson.

Inspired by Greene & Greene, cloud-lift patterns in mahogany surround a Craftsman fireplace (photo below). Designed by Berkeley Mills. Built by DeSmidt Builders. Photo by Gene DeSmidt.

INDEX

The articles in this book originally appeared in *Fine Homebuilding* magazine. The date of first publication, issue number and page numbers for each article are given at right.